W9-CMB-376

TROPICAL GEOMORPHOLOGY

Focal Problems in Geography Series

Published

DAVID GRIGG: The Harsh Lands: A Study in Agricultural Development

ALAN HAY: Transport for the Space Economy: A Geographical Study

MICHAEL F. THOMAS: Tropical Geomorphology: A Study of Weathering and Landform Development in Warm Climates

Titles in preparation include

J. T. COPPOCK and B. S. DUFFIELD: Recreation in the Countryside: A Spatial Analysis

BARRY J. GARNER: Linkages Within Cities

Tropical Geomorphology

A Study of Weathering and Landform Development in Warm Climates

MICHAEL F. THOMAS

Department of Geography
University of St Andrews

MACMILLAN

First published 1974 by
THE MACMILLAN PRESS LTD
London and Basingstoke
Associated companies in New York Dublin
Melbourne Johannesburg and Madras

SBN 333 03313 2 (hard cover)
333 11772 7 (paper cover)

Printed in Great Britain by
THE ANCHOR PRESS LTD
Tiptree, Essex

Distributed in the United States and Canada
by Halsted Press, a Division of
John Wiley & Sons, Inc., New York

Library of Congress Catalog Card No.: 73–13428

To Anne

Contents

List of Plates

Between pages 148 and 149

Preface

In the preparation of this book I have drawn upon the help of many people and institutions. Many of these have no direct connection with the writing of the book, but without their assistance, which made the necessary field experience possible, this study could not have been written.

Financial assistance and material help has been provided by the Carnegie Institution, which financed work in West Africa in 1967, and in Papua and New Guinea in 1970; by the University of Ibadan, which consistently supported my fieldwork in Nigeria; by the University of St Andrews, which has supported work overseas, and the University of New South Wales, which assisted in undertaking field work in Australia. In addition I have received help in the field from the following: Associated Tin Mines of Nigeria Limited; Sierra Leone Development Company; Geological Survey of Nigeria; Australian National University; University of Queensland; and the James Cook University of North Queensland.

Among very many individuals who have helped me in various ways during field work in the tropics, I am particularly grateful to the following: in Nigeria Dr N. K. Grant, Dr D. C. Ledger, Professor R. P. Moss, Dr H. Vine and Dr G. A. Worrall, all formerly of the University of Ibadan; Eliezor, who drove hundreds of miles throughout Nigeria on my behalf, and to Dr R. A. Pullan and Dr M. B. Thorp, both formerly of the Ahmadu Bello University, Zaria. In Sierra Leone I was greatly helped by Dr and Mrs G. J. Williams; in Singapore by Dr and Mrs B. St J. Swan; in Papua by Mr and Mrs M. Browne; in Australia by Professor J. A. Mabbutt and by Dr and Mrs D. Hopley. I should also like to acknowledge the interest taken and stimulus given to my work by Professor J. N. Jennings, Professor L. C. King, Professor I. Douglas, Dr C. D. Ollier, Dr F. Ojany, Dr H. M. Churchward, Dr M. J. Selby, Dr B. P. Ruxton, and Mr M. Finkl, Jr, all of whom have taken time to show me things I could not have seen on my own. I should also like to thank Professor J. C. Pugh, who guided some of my early work in the tropics. Others whom I have not had the good fortune to accompany in the field have contributed immeasurably through the stimulus of their work and ideas concerning the development of tropical landscapes. Their names appear in the bibliography, and I can only hope that I have correctly interpreted their work.

The preparation of this book was made possible by the skill and patience of Mrs Elizabeth Kerr, who typed most of the manuscript; Mr J. P. Clarke and Mr J. Davie, who drew all the maps and diagrams, and Mr P. Adamson, who printed my own photographs. Most of all I am grateful to my wife, Anne, who has had to live with this book for too long. The shortcomings of the book are however all my own.

The author and publisher would also like to thank the following publishers for permission to reproduce copyright tables and figures: Academic Press, New York (table 8); American Institute of Mining and Metallurgy (figure 8); Centre de Documentations Universitaires, Paris (figures 19 and 39); The Clarendon Press, Oxford (table 5 and figure 17); Department of Scientific and Industrial Research, N.Z. (table 3); Ferd. Dummlers Verlag, Bonn, and Geographisches Institut, Bonn (figure 29); Gebrüder Borntraeger and the Zeitschrift für Geomorphologie (figures 3, 27, 32, 33, 34, 36, 40); Göttinger Bodenkundliche Berichte (table 16); Institute of British Geographers (figures 20–24; 46); Institute of Civil Engineers, London (figure 10 and table 4); Institut Geographique Nationale, Paris (figure 35); McGraw-Hill Book Company (figure 4); Masson et Cie, Paris (figure 2), and Springer, New York (figure 12); Methuen and Company, London (table 14); Museum National d'Histoire Naturelle, Paris (figure 45); Oliver and Boyd, Edinburgh (figure 1); Royal Geographical Society (figures 16 and 28). In addition the following authors have granted permission to reproduce their material: A. Aubreville (figure 45); L. Berry (figures 19, 39, table 9); H. M. Churchward (figure 44 and table 15); D. Craig and F. C. Loughnan (figure 7); G. H. Dury (figure 11); T. Feininger (table 14); M. Fieldes and L. D. Swindale (table 3); F. Fournier (figure 26); H. Jenny (figure 9); P. Lumb and E. C. Ruddock (figure 10); R. Maignien (figures 13, 14, 15); H. L. Ong, V. E. Swanson and R. E. Bisque (figure 6); G. Pedro (figure 2); B. P. Ruxton (figures 19, 25 and 39); S. B. St. C. Swan (table 11); UNESCO (figures 4, 6, 8). I am also grateful to A. Young for permission to quote unpublished results from experiments on slope wash under rain forest in West Malaysia and Singapore.

St Andrews 1974 MICHAEL F. THOMAS

Introduction

The aim of this book is to establish a basis for the study of land-surfaces that have evolved by the operations of a complex group of processes associated with chemical weathering, mass movement and surface water flow. The influences of frost, snow and ice on the one hand and of wind on the other are specifically excluded, as are coastal processes. The type areas for such an analysis lie within, or border, the inter-tropical zone, although it can be argued that most of the world's land area beyond the glacial and periglacial zones is primarily developed by such processes. However, the present-day influence of frost within the temperate and cold climatic zones, together with the fundamental effects of the Quaternary climatic oscillations on these areas, appear cogent reasons for studying first the landscapes of the tropics and sub-tropics. Beyond this, the principles adduced for such a study can be utilised to assist in the understanding of the landforms of higher latitudes, but this latter task cannot be pursued very far in the present context. This book is not therefore intended as an apologia for a particular form of climatic geomorphology, but rather as an enquiry into the conditions of landform development within a specific range of bioclimatic conditions.

Because of limitations of space it is necessary to assume that the reader is familiar with the character and range of tropical climates and their associated vegetation communities. However, while no systematic account of these fundamental environmental conditions can be given here, their importance is recognised and referred to wherever appropriate.

Diastrophism and climate in landform development

It has long proved difficult to demonstrate in detail and with precision the effects of crustal upheaval upon landforms, although they are measured indirectly in the study of available relief, depth of dissection and slope conditions within the landscape. Nevertheless, there is much evidence to indicate that a high proportion of the tropics is underlain by old and relatively stable landmasses, and tectonically mobile zones are found mainly in Indonesia, Central America and the Caribbean. This situation leads inevitably to an emphasis upon the development of landforms on 'old lands' (Waters 1957), comprised mainly of crystalline, igneous and metamorphic rocks, and often to the formulation of models of weathering and landform development within areas of granite or granitoid rocks.

It can be argued that the forms and deposits found on these tropical cratons exhibit features which are not specifically tropical, but are rather functions of the duration of weathering and the persistence of a low rate of denudation consequent upon prolonged crustal stability. Conversely, these conditions offer a unique opportunity for the study of landforms developed over an extended period, under the environmental conditions specified above. In fact, the absence of major crustal upheaval and of fundamental bioclimatic disturbance in certain areas of the tropics, during a time span of 10^5–10^6 years, has led to a singular continuity of weathering and erosion, and it is possible to reason that in these areas, if anywhere, will be found the landforms and deposits which may be associated with broadly tropical conditions. But this does not, of course, indicate that they may not also be found elsewhere. Furthermore, the crystalline, silicate rocks offer an ideal basis for an enquiry into the separate effects of chemical decay and mechanical abrasion, on account of the fundamental chemical alteration which these rocks undergo during weathering, and of the major changes of erosional mobility that accompany weathering.

In tectonically mobile areas, weathering penetration is limited by high rates of surface denudation, and Ruxton (1967) has suggested that differentiation of landforms on the basis of climate under these conditions may be very limited indeed. However, few areas of the earth's crust have remained strictly stable over long periods, and the effects of uplift, warping, and rifting accompanied by continental drift are by no means absent from the ancient landsurfaces. But except along narrow zones of rifting or monoclinical warping, associated with the fragmentation of larger crustal plates such as Gondwanaland, most of these changes have produced a subtle shift in the denudation system, and not a replacement of one landsurface by another through the *direct* intervention of endogenic forces. Bioclimatic changes appear also to have operated in a comparable fashion, leading mainly to shifts in the balance of existing process. The intrusion of new processes has occurred only in semi-arid areas, where eolian conditions have periodically been active.

Despite doubts concerning the identification of 'tropical' landforms, the products of weathering clearly vary in character and amount according to climate (Strakhov 1967; Pedro 1968; and figures 1 and 2). However, it is important to recognise that these variations do not arise from the operation of different processes, but from the manner in which 'universal' processes operate under different environmental conditions. A similar argument may be advanced concerning landform development, for although water flow obeys universal laws, it is false to argue that all fluvially eroded landscapes must be identical. Nevertheless, the recognition of morphogenetic regions has usually been based upon deductive argument (Büdel 1957, 1968; and figure 3), lacking in an adequate basis of empirical data (Stoddart 1969). Tricart (1972) continues to argue strongly for a distinctive, tropical morphogenetic zone, but such a view rests, as Stoddart (1969) has emphasised, mainly upon the identification of apparently distinctive individual landforms, about which we as yet know comparatively little, rather than upon the general morphometry of the terrain. Certain of these forms, such as the domed inselbergs or bornhardts, have aroused great controversy when used for this purpose (King 1957). But although these forms will be discussed in detail later, no attempt will be made to set strict climatic limits for their occurrence or development.

The landsurface and the ecosystem

The part which deep chemical weathering plays in the development of tropical landforms has emphasised the need to look below the actual land surface for any comprehensive account of the landform. Recent developments in morphometry, particularly in the fields of hillslope analysis and drainage composition, have taken little account of this requirement, and as a result have contributed less to the understanding of tropical geomorphology than work in pedology and soil mechanics. In fact, the study of surface forms as though they had a tangible reality apart from the materials over which they are developed must be of limited value, and it commonly derives

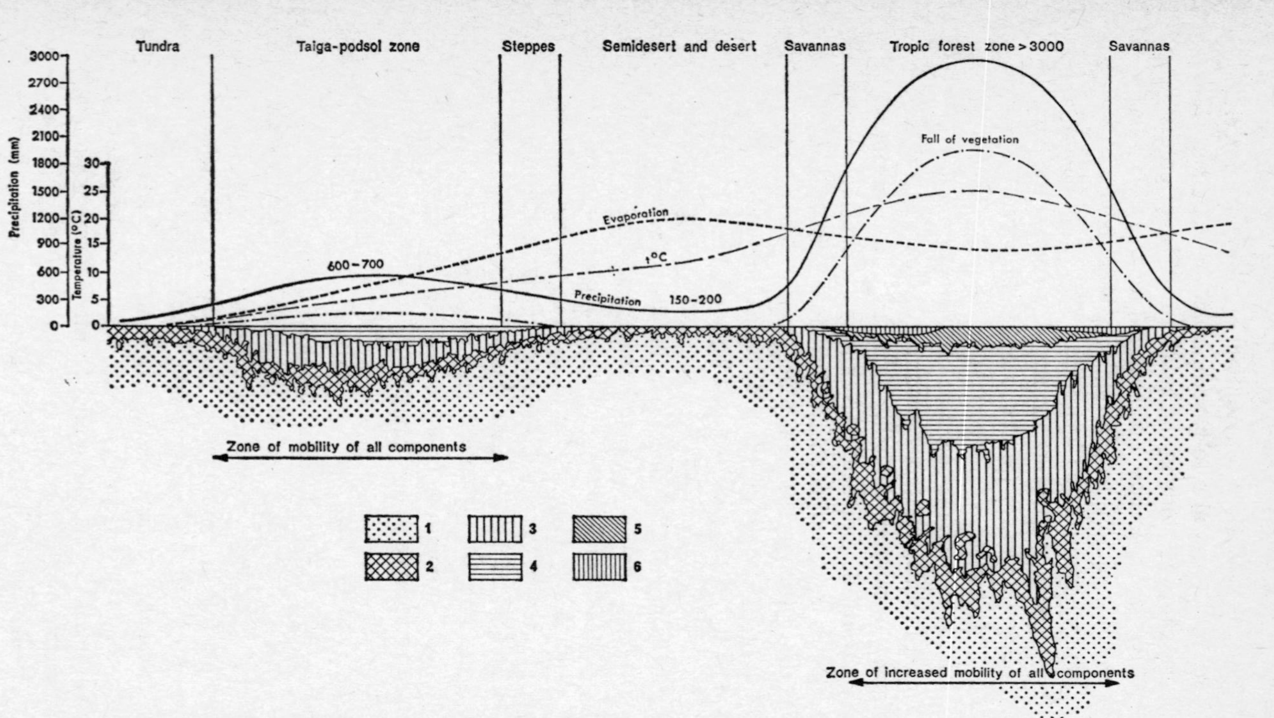

Figure 1 Sketch of the formation of weathering mantles in areas that are tectonically inactive (from Strakhov 1967)

1. Fresh rock; 2. Zone of gruss eluvium, little altered chemically; 3. Hydromica-montmorillonite beidellite zone; 4. Kaolinite zone; 5. Ocher, Al_2O_3; 6. Soil armour, $Fe_2O_3 + Al_2O_3$

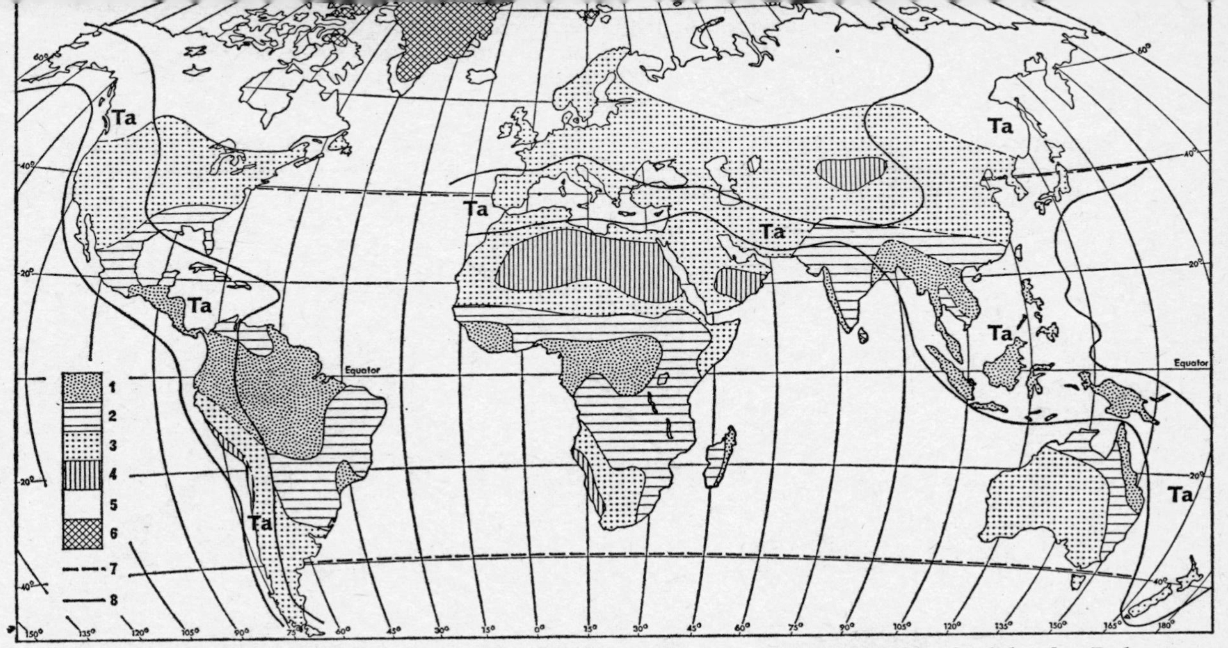

Figure 2 Distribution of the principle types of weathering on the earth's surface (mainly after Pedro 1968, with additions from Strakhov 1967)

1. Zone of allitisation (kaolinite plus gibbsite); 2. Zone of monosiallitisation (kaolinite); 3. Zone of bisiallitisation (2:1 lattice clays–montmorillonite, illite); 4. Hyper-arid zone without significant chemical weathering; 5. Zone of podsolisation; 6. Ice-covered areas; 7. Approximate limit of red weathering crusts (after Strakhov); 8. Extent of tectonically active areas (TA) within which 'climatic' weathering types are modified.

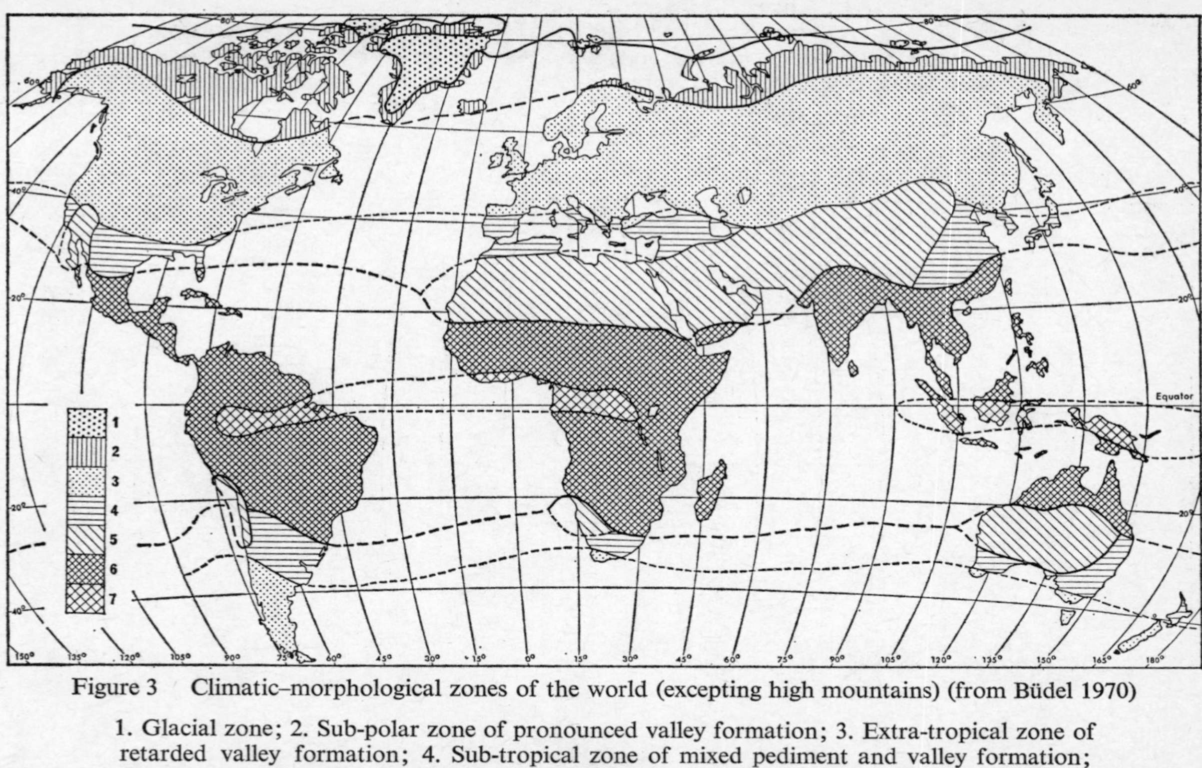

Figure 3 Climatic–morphological zones of the world (excepting high mountains) (from Büdel 1970)

1. Glacial zone; 2. Sub-polar zone of pronounced valley formation; 3. Extra-tropical zone of retarded valley formation; 4. Sub-tropical zone of mixed pediment and valley formation; 5. Arid zone of surface building; 6. Outer tropical zone of pronounced planation surface formation; 7. Inner tropical zone of partial planation surface formation.

from the manipulative convenience of map and aerial photographic analysis, as an alternative to painstaking field enquiry.

Consideration of the landform as a surface without thickness also neglects the position of the landsurface within the ecosystem, where it forms an interface between biosphere and atmosphere above, and lithosphere below. Because in reality these materials overlap in the soil and regolith, which are parts of the ecosystem, the study of landform must be concerned with a multi-layer situation, involving bedrock, sedentary regolith and transported mantle(s), as well as horizon differentiation within soil and weathering profiles. The presence or absence of such layers, or of characteristic horizons within them, thus mark important variations in landsurface conditions that may not in all cases be marked by corresponding changes in slope. Therefore, components of the landform which may be described as facets (Beckett and Webster 1965), are commonly differentiated according to the nature of sub-surface materials, slope and drainage conditions.

The denudation system which maintains or modifies the landsurface by removal or accumulation of material must therefore be regarded as a sub-system within the ecosystem. Indeed Douglas (1969, p. 2) suggests that, 'to a large extent the ecosystem and the denudation system, at a given point in space and time, are identical, and can be discussed in terms of the dynamics of ecosystems'. Such a view makes clear the complexity of geomorphology, and the difficulties of analysing landform development at scales which would require an understanding of several interacting, and individually complex, community ecosystems. This study cannot claim to attempt such a task, partly because the necessary data are not available. Nevertheless this viewpoint is valuable, for it emphasises the nature of the landsurface as a dependent variable within a complex system, and shifts attention away from some of the more arid controversies concerning slope evolution.

Equilibrium concepts and time-dependent landforms

If landform and soil are viewed as open systems, it is possible to consider both in terms of equilibria and the attainment of steady state conditions, in which evolutionary concepts are out of place. Nikiforoff (1949, 1959) explicitly challenged the evolutionary view of soil development, and considered that the biological analogy of ageing as applied to soils is in itself inappropriate, and misleading in the correct interpretation of field evidence. Hack (1960) and Chorley (1962) have offered a comparable view of landform development, thus challenging the basic tenets of W. M. Davis's concept of the cycle of erosion. This equilibrium view of the soil and landform allows for the lowering of the landform and the soil profile with time, but without identifiable changes in the form of either.

Nikiforoff (1949, 1959) considered that the soil profile might be lowered by erosion of the landsurface and simultaneous sinking of the horizon of decomposition within the rock. As a corollary, soil horizons must be capable of reversible changes, permitting the transformation of C horizons to B horizons, and the latter to A horizons. This view of the soil is relevant to certain questions surrounding laterite formation, and conforms with the concepts of Trendall (1962) and De Swardt (1964) concerning the development through time of old landsurfaces.

However, Schumm and Lichty (1965, p. 110) have pointed out that 'although landforms are physical systems . . . they are also analogous to organisms because they are systems influenced by history'. These authors attempt to reconcile the systematic and evolutionary approaches to the landform by recognising a conventional sub-division of the time scale of landform development. In this, evolutionary aspects of the landform are seen within 'cyclic' time of geological dimensions, while equilibrium forms persist as components of the landscape over short periods of 'graded' time of measurable duration. Particular elements of the landform system may achieve a true steady state only within very short periods. Schumm and Lichty (1965, p. 118) conclude that 'the more specific we become the shorter is the time span with which we can deal, and the smaller is the space we can consider, conversely in dealing with geologic time we generalise'.

Such concepts affect landform analysis in fundamental ways. Thus, if the equilibrium view is strictly followed, the spatial differentiation of landforms results mainly from differences in bedrock (Hack 1960), whereas, in the conventional methodology of geomorphology, forms arranged in space are commonly taken to represent stages of development through time. This follows a deductive hypothesis of landform development, comparable with, though not necessarily the same as, that of W. M. Davis. It is often based on comparative morphology (in soil science also) which involves hazardous reasoning: first, because it assumes that all landforms in an area have passed, or will pass, through the same sequence of changes, and, second, because it assumes that only via a particular evolution can any individual landform arise. These premises deny the possibility of geological control (Hack 1960), and also the principle of 'equifinality' (Bertallanffy 1950) or 'convergence' (Cunningham 1969), whereby a single form may arise via different sequences and from separate origins. This problem was clearly posed by White (1945) in a perceptive study of bornhardt domes in the south-eastern piedmont of the United States.

In the study of stochastic processes, the replacement of space averages for time averages may be justified, but attempts in this field relate only to systems in a steady state which respond rapidly to changes in external variables. In natural landscapes, different aspects of the land complex react to such changes at widely varying rates. Ruxton (1968c) has pointed out that, while some features such as river channels are maintained close to a state of dynamic equilibrium, others such as deep weathering profiles respond so slowly that they remain almost indefinitely in a state of disequilibrium. Inevitably such a situation must give rise to 'inheritance' of forms and deposits within the dynamic systems of the present, and to what Ruxton describes as 'disorder' in the landscape.

Theoretical studies of the landform suggest that the persistence of time-independent forms is improbable (Schumm 1963; Ahnert 1970), and that crustal upheaval in any one area is sporadic, leading to rapid revival of relief and followed by longer periods of quiescence, during which the landforms 'age'. This Davisian concept, even when shed of all detail, is one of decay (Ruxton 1968c) or increasing entropy (Chorley 1962), and implies that in the absence of diastrophism the landform will experience a progressive reduction of differentiation, leading towards a 'peneplain'.

On the other hand Nikiforoff (1959) considered the energy system of the soil in terms of 'ceaseless excitation' by solar radiation which enables the soil system to function continuously through geological time. This concept can be applied to the observation that weathering profiles of great age are commonly strongly differentiated into characteristic horizons which do not suggest a high level of entropy in the system. It is possible therefore that landforms and soils respond in different ways to energy inputs into the denudation system.

However, when these concepts are viewed in terms of field evidence, it becomes necessary to avoid imposing a rigid conceptual framework upon landform studies. Ruxton (1968c) points out that the landscape is comprised of 'non-regular, non-random systems' which arise from the 'multicomplexity of process and inheritance'. Additionally, it is necessary to recognise that virtually all natural systems experience cyclic changes or pulses that vary in their effects upon the landscape in time and space. The periodicity of such pulses varies, but many interrupt long-term cycles of decay, and some disrupt equilibrium states in the landform so that their persistence in time is unlikely.

The natural landscape thus exhibits a mosaic of facets of varying form and surface materials. In this study an attempt will be made to recognise some of the principal forms which these take in the tropical landscape, and to interpret these in terms of the processes which appear to be responsible for them. In such an attempt, differences between landsurfaces on bedrock, on regolith and on transported mantles are considered to be as fundamental as slopes of varying declivity. Similarly, the effects of differential erosion (Hack 1960) must be considered alongside the results of sequential development.

Part One

Geomorphic Processes in the Tropics

1 Weathering Processes and Products

Recent studies do not support the concept of a tropical zone within which an exclusive group of distinctive weathering processes occurs. Fripiat and Herbillon (1971, p. 21) state clearly that they 'do not believe that the transformation sequences or the synthesis mechanisms observed in tropical regions are essentially different from those occurring under other climatic conditions.' Nevertheless, the characteristics of the weathering products found in tropical regoliths are commonly found to be distinctive, and Pedro (1968) has recognised zonal weathering types according to the dominance of particular clay minerals and the behaviour of iron (figure 3). Furthermore, depths of rock decay are frequently held to be greater in the tropics than elsewhere. This view must also be accepted with caution, because deep weathering as a phenomenon is not confined to the tropics, and also because within the tropical zone itself characteristic weathering depths vary greatly, particularly with the humidity of climate, but also in response to many other factors (table 1).

Ruxton and Berry (1961a) suggest that beneath landsurfaces of similar age, expected depths of decay may be 30 m in the humid tropics; 25 m in the wetter savannas, falling to 6 m in the drier savannas and to less than 3 m in the arid zone. Deep weathering is in fact common within the entire extra-glacial zone. Feininger (1971), and Demek (1964) suggest that depths of 20 m may be characteristic of granites in Bohemia. Lesser depths, possibly about 6 or 8 m, are more common, but much deeper profiles are found locally. Although much of this deep weathering has been attributed to tropical conditions prevailing during the Tertiary in northern and central Europe, such a view may not be necessary.

Nevertheless, widespread deep weathering, together with certain regolith characteristics, are considered here to be fundamental to any enquiry into tropical geomorphology. Some of the factors contributing to the rate and course of weathering are listed in table 1.

TABLE 1

MAJOR FACTORS AFFECTIN DEPTHS OF WEATHERING

Climatic factors	Temperature: high temperatures increase the rate of endothermic chemical reactions.
	Precipitation: high precipitation increases the availability of the principal reagent in weathering processes: water.
Biotic factors	Vegetation cover: a dense forest canopy protects the surface from wash processes and provides organic acids which are capable of mobilising certain rock minerals, especially iron by chelation. Conversely, open vegetation of the savanna type favours immobilisation of iron and favours surface runoff.
Geomorphic factors	Landsurface stability: weathering penetration is favoured by a low rate of surface denudation prevailing on gentle slopes.
	Age of landsurface: prolonged stability (persistence of ancient landsurfaces) allows deep profiles to develop.
Site factors	Free drainage: elevated sites promote downward movement and frequent renewal of groundwater essential to rapid decomposition of rocks.
	Reception zones: sites experiencing convergent runoff may have increased water supply, but this may combine with poor site drainage.
Geologic factors	Rock type: presence of minerals particularly susceptible to alteration increases rate of weathering penetration and may promote early disaggregation of rock.
	Rock texture: rock texture affects behaviour under weathering attack. Crystalline rocks of coarse texture disaggregate more rapidly than fine textured rocks which may undergo more rapid alteration. Texture in sedimentary rocks affects permeability and rate of weathering penetration.
	Rock fissility: faults, joints and fractured grain boundaries promote weathering penetration, especially in crystalline rocks.
	Hydrothermal alteration: rocks previously subject to varying forms of hydrothermal activity may become additionally susceptible to groundwater weathering on exposure.

Chronologic factors Climatic change: variations of climate and vegetation with time alter the balance of weathering and erosion. Pluvial conditions in the arid zone during the Tertiary and Pleistocene have led to the presence of relict deep weathering.

Tectonic change: variations of crustal stability affect landsurface stability and the period available for weathering penetration.

From this table some very general predictions may be made, but the large number of variables involved makes detailed prediction very difficult. However, it is clear that the inter-tropical zone is one within which many of these factors operate to favour deep chemical decay of rocks. Large areas of high temperature, and moderate to high rainfall promote the weathering processes, and outside of south east Asia and parts of the Andes and Caribbean, the tropical zone is dominated by ancient crystalline massifs and associated sedimentary basins. Stable landsurfaces of considerable antiquity thus provide a geological and geomorphic basis favouring deep penetration of weathering, and thorough decomposition of rocks. Weathering processes are most favoured in the humid tropics, where surface denudation is not particularly rapid (Fournier 1962; Douglas 1969). But in the drier savannas, where weathering is inhibited by seasonal drought, and surface erosion is rapid, (Fournier 1962; Schumm 1963) many deep profiles may be relics of earlier, more humid periods. Although depth is often taken as the most significant geomorphic factor associated with the weathering phenomena, the nature of weathering products is a more accurate guide to the effects of climate, and may also be of importance to an understanding of the behaviour of regoliths to subsequent surface denudation.

The majority of rock-forming minerals are silicates, and this discussion therefore turns upon the reaction of silicate minerals to weathering environments. Our knowledge of the natural environments of weathering is still scanty, and this is not surprising, since many such environments exist tens of metres below the ground surface. They also involve the complexities of organic matter, of wetting and drying and of mutual interaction of various minerals and weathering products.

Most of the chemical reactions which take place, in particular hydrolysis which is the most important, are *endothermic* and their rate increases exponentially with temperature. A figure of 2 or 3 times is usually given for increases in the rate of reaction for every 10°C rise in temperature. This is known as Van't Hoff's rule. Since soil water temperatures, as between the temperate and tropical zones, are likely to differ on average by 15°C or even 20°C it can be argued on grounds of temperature effects alone, that tropical weathering rates will exceed those of the cool temperate areas by at least 4 times.

However, temperature alone does not control the rate of weathering, and quite apart from the topographic factor, the supply of groundwater and of organic matter are of particular importance. In the availability of these materials contrasts within the tropical zone may be greater than within the temperate latitudes. However, Strakhov (1967) has pointed out that in the humid tropics temperatures vary between 24–26°C, rainfall between 1200–3000 mm, and organic matter from the rain forest is supplied at an annual rate of 100–200 tons per hectare. By contrast, the cool temperate climates experience annual temperatures below 10°C, rainfalls of from 300 to 700 mm and, within the taiga zone, an annual organic increment of 8–10 tons per acre. Strakhov claims that the greater leaching within the humid tropics will increase weathering rates 7–14 times, and that when all factors are accounted for the differential may be as much as 20–40 times. These are extreme figures and such comparisons may be of little value, because precise definition of the weathering environments being compared is generally lacking.

In many respects it is the availability of water as the principal chemical reagent involved in weathering processes that becomes most critical, especially within the tropical zone. Chemical reactions will cease if concentrations of dissolved salts rise towards equilibrium conditions (Le Chatelier's Law), and therefore such equilibria must continually be disturbed if weathering is to continue. This requires the renewal of groundwater and/or the removal of solutes. Furthermore, access to fresh rock is eventually inhibited by the presence of the residual products of weathering, especially clays which may be poorly permeable. Such factors probably impose limits to possible depths to which groundwater weathering may penetrate. Dixey (1931) also pointed out that at great depths the confining pressures would exclude further groundwater penetration. In practice, removal of superficial material eventually outpaces further weathering penetration, and so limits actual depths of weathering.

The products of weathering are further influenced by delicate adjustments within the system. These hinge mainly upon the pH (concentration of H ions) of the groundwater, the potential for oxidation or reduction (redox potential or Eh), and the presence of organic matter which promotes the formation of ring-structures containing metal cations (chelation). To explain these further some account of the weathering processes must be given, but the reader is referred to standard texts for a full treatment of this subject (Reiche 1950; Keller 1957; Krauskopf 1959; Loughnan 1969; Ollier 1969; Carroll 1970).

The weathering of silicate minerals

Silicate minerals are built from crystal lattices which are comprised basically of silica tetrahedra within which the silicon (Si^{4+}) atom shares a single bond with each of four oxygen (O^{2+}) atoms. Such tetrahedra can exist singly (as in quartz) or in chains and sheets. By ionic substitution silicon atoms within these sheets may be replaced by metal cations such as aluminium (Al^{3+}), iron (Fe^{3+} or Fe^{2+}), magnesium (Mg^{2+}), calcium (Ca^{2+}), sodium (Na^{+}) and potassium (K^{+}). In this way many silicate minerals are formed and they comprise the bulk of most common rocks. The substitution of the metal cations takes place within rock magmas and is facilitated by the high energy conditions associated with very high temperatures. The bonds formed between the ions may be weak at low temperatures, in the presence of H^{+} and OH^{-} ions from water. It is found that the higher the energy of formation, which may be seen in terms of the temperature of crystallisation, the easier the disintegration of the mineral by hydrolysis.

This can also be interpreted in terms of the $Si:O_2$ ratio. Where substitution of Si by other cations has been great the ratio is low, but where it has been less the ratio is higher. The higher the ratio the less susceptible to weathering is the mineral.

Two groups of silicate minerals can be identified: the ferro-magnesian (mafic) minerals, typified by olivine ($[MgFe]_2SiO_4$) in which isolated silica tetrahedra are linked by Fe^{2+} and Mg^{2+}, and the feldspars within which Al^{3+} substitution is accompanied by the acquisition of Ca^{2+}, K^{+} or Na^{+}. In each group a series can be constructed illustrating the range of susceptibility.

This series (after Polynov (1937) and Goldich (1938)) corresponds closely with the order of crystallisation from a silicate melt recognised as the Bowen Series:

Olivine
1 : 4* Pyroxene (Augite)
1 : 3 Hornblende
4 : 11 Biotite
2 : 5 Orthoclase–Quartz–Muscovite
3 : 8

Calcic-Plagioclase Sodic-plagioclase
1 : 4 3 : 8

Least stable ————————————————→ *Most stable*

* Ratios are given for $Si : O_2$.

The ferro–magnesian minerals form a discontinuous series, but the feldspars form a continuous series with almost infinite gradations of composition. The reasons for this sequence are not fully understood, but in olivine for instance the magnesium and ferrous ions are held on the edges of isolated tetrahedra (Loughnan 1969) and are easily detached. However, although the alkali feldspars contain mobile ions of potassium and sodium these are held in a framework of linked tetrahedra which hinders their escape. Such contrasts in crystal structure are an essential part of the understanding of weathering processes, but cannot be pursued here.

Other factors also intervene. For instance biotite frequently weathers more rapidly than hornblende, and commonly leads to the early disaggregation of granites (Eggler *et al.* 1969). This has to do with the usually larger crystal size of biotite and the tendency for fractured grains to occur in the fresh rock.

It is difficult to rank rocks in a similar manner to minerals for a variety of reasons. First, rocks have no precise chemical formula, and two rocks similar in hand specimen may contain slightly varying proportions of the same minerals, or may differ from each other in the addition of a minor new mineral, in such a way that reactions to weathering processes will vary quite markedly. Rocks also vary in texture, in the strength of bonding, and in internal stresses and fractures in ways that are extremely difficult to isolate.

Nevertheless a guide to rock weathering can be taken from average compositions of common rocks and Rougerie (1960) offers a sequence:

MOST STABLE
↑
Pegmatite
Porphyritic granite
Alkaline granite, mica schist and quartzite
calc-alkaline granites with two micas, gneiss
granodiorite
diorite, gabbro
microgranite, dolerite, amphibolite, pyroxenite
microdiorite, microgabbro and metamorphic schists
green rocks
↓
LEAST STABLE

But Rougerie did not find this ordering an accurate guide to relief development, and we must conclude that, only at a most general level can rocks be ranked into a weathering sequence. Nevertheless, within the granite group of rocks it is clear that the nature of the feldspars and the possession of biotite greatly influence the susceptibility to weathering. Questions of texture and jointing will be considered in a later section.

The course of weathering or breakdown of silicate minerals depends largely upon the process of hydrolysis, or the activity of ionised water. In general, the H^+ ions replace the metal cations in the silicate lattice and the OH^- ions combine with these cations to form soluble products which may be evacuated from the system. But the character of the weathering products will depend upon the weathering environment, because mobility amongst the common cations in the silicate rocks varies widely and is affected by the pH of the groundwater, and in the case of iron by Eh as well. The sequence of mobilities amongst metal cations in rocks has been established in several studies, including those by Anderson and Hawkes (1958), Dennen and Anderson (1962), Harris and Adams (1966) and Loughman (1969). Their results may be summarised:

(1) Most mobile Ca^{2+}, Na^+, (Mg^{2+}) (K^+)
(2) Intermediate K^+, Mg^{2+}, Si^{4+}, Fe^{2+}
(3) Least mobile Fe^{3+}, Al^{3+}

Both magnesium and potassium are readily mobilised, but they tend to recombine to form clay minerals, though under strong leaching magnesium is normally lost in solution, as part of group 1.

Silicon has a low but significant mobility at most pH levels (figure 4) when it occurs as amorphous silica, or is released from silicate minerals. As quartz it has a solubility less than one tenth these values (Krauskopf 1959). Iron mobility depends on the Eh or redox potential: in the oxidised ferric state it is immobile, but in the reduced ferrous form may go into solution readily at common soil pH values, usually as the bicarbonate (figure 5).

The behaviour of the three groups can thus be predicted: the most

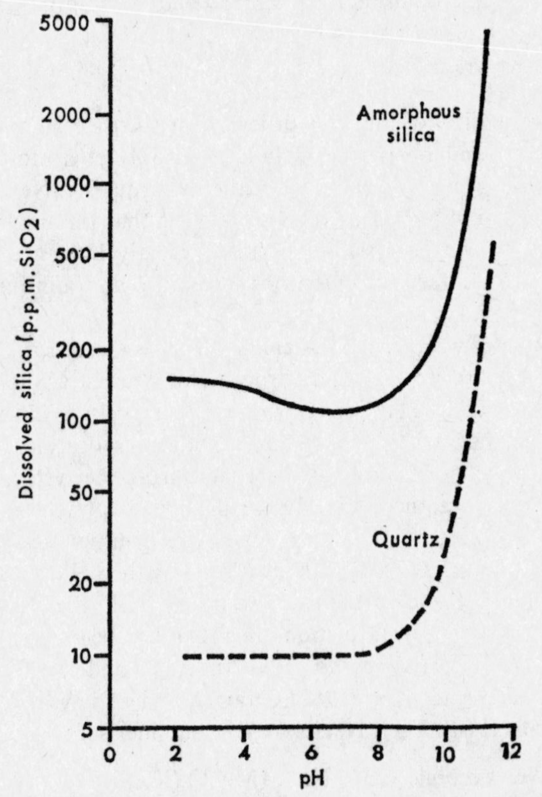

Figure 4 Solubility of silica at 25°C (from Krauskopf 1959)

The solid line shows the variation in solubility of amorphous silica with pH, as determined experimentally. The lower dashed line is the calculated solubility of quartz, based on the approximately known solubility of 10 p.p.m. SiO_2 in neutral and acid conditions.

mobile elements are usually lost to the site of weathering in solution, the elements of intermediate mobility tend to remain in solution for short periods, often under critical environmental conditions,

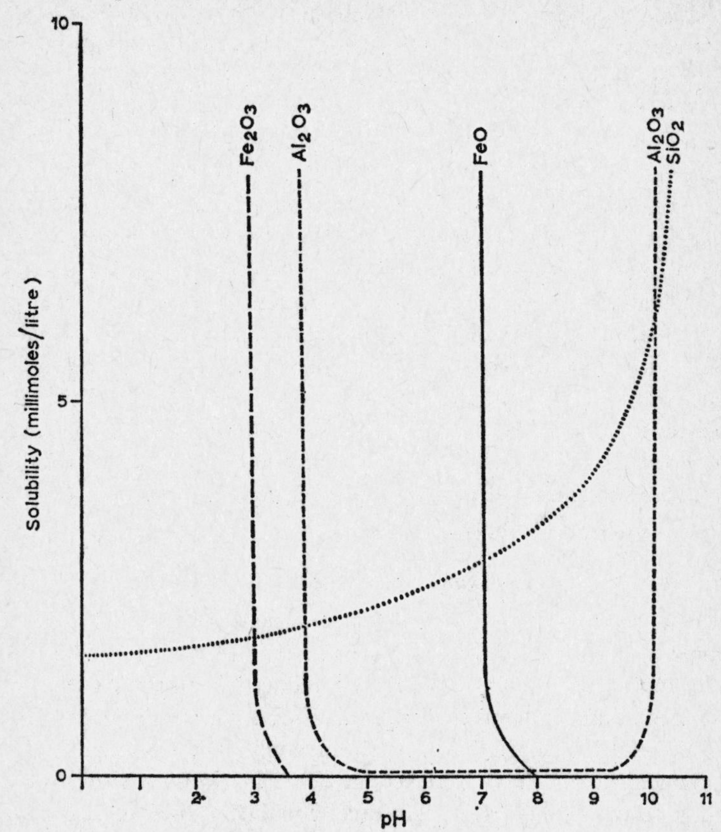

Figure 5 Variations in solubility of iron and aluminium oxides and silica with pH.

and are commonly redeposited in new minerals (often clays) close to the site of weathering. The least mobile of the common elements: aluminium and iron in the ferric state, generally remain as residual products of weathering, usually as sesquioxides.

In the formation of tropical regoliths it is the pH of the groundwater and the presence or absence of air, together with the high soil water temperatures and presence of organic acids, which are of greatest importance. In figure 5 solubilities of the three least mobile elements in the system are related to the pH of the soil solution. From this it can be seen that silica is moderately soluble throughout the common range of soil pH but that its solubility rises steeply under alkaline conditions. With a pH above 9 both silica and alumina are readily removed in solution. Under acid conditions with pH of 4 or less, alumina once again becomes soluble, but silica is much less mobile. Iron in its reduced form (FeO) is highly mobile under acid or neutral conditions, but is precipitated at a pH of 8 or above. Iron in its oxidised (Fe_2O_3) form is only soluble in highly acid conditions, pH 3 or less, which are extremely rare in tropical soils, so that for practical purposes the oxidation of ferrous iron compounds leads to their precipitation. As the reduced form of iron is unstable, it is readily oxidised in free draining, aerated soil environments. There is, however, a complicating factor, and this is the readiness which iron displays for organic complexing (chelation), whereby it goes into a soluble ring structure with organic compounds. This only occurs in a soil environment rich in organic matter. It is clearly necessary to look carefully at the pH of weathering environments in order to determine the course of weathering reactions and in particular their stable products.

Weathering and groundwater conditions

Recent work has demolished older ideas that tropical rainfall contains a high concentration of CO_2. It is clear that equilibrium concentrations of CO_2 in water are reduced by increases in temperature. Similar claims that tropical rainfall contains large amounts of dissolved nitrogen have not been confirmed. Rougerie (1960) estimated that rainfall received at Abidjan (Ivory Coast) has a range from pH 5·1–7·4, so that exceptional conditions of acidity or alkalinity cannot be associated with rainfall as such. But the pH of water is affected by its reactions with organic matter and with rock minerals. Hydrolysis itself can be demonstrated by testing what is known as the 'abrasion pH' of minerals, by grinding them in water. The pH of the resultant suspension characterises individual minerals and aids their recognition (Stevens and Carron 1948). The high abrasion pH value of common silicate minerals is due to the replacement of metal cations by hydrogen ions from the water. The release of the cations raises the pH of the solution, particularly in the case of the alkalis and alkali earths (Ca, Na, K, Mg). The actual values of pH obtained closely follow the weathering sequence just demonstrated, and range from pH 10–11 for olivine, hornblende and augite, pH 8–9 for biotite and plagioclase, to pH 6–7 for muscovite and quartz. This series reflects the relative strength of ionisation of metal cations in silicate rocks, and their tendency to become replaced by hydrogen ions in ionised water. The very high pH values obtained in the laboratory are not often attained in natural ground water, although Bakker (1960) found that water evaporating from bare granite surfaces in the tropics became alkaline, the pH rising to 8 or 9.

Within freely drained soils and regoliths, however, leaching of cations combined with OH^- leaves an excess of H^+ giving pH values commonly in the range 5–7. Acidity is also promoted by the liberation of CO^2 by organic matter, and by the presence of humic and fulvic (organic) acids, while leaf extracts commonly give pH values of 4–5 (Carroll 1970, table 31). Very low pH values have been recorded from swamp environments, and in Guyana, Bleackley (1964) gives figures of 2–3 for peats: low enough to take ferric iron into solution.

However, the widespread development of groundwater acidity (see table 2) requires a more general explanation. Water molecules belonging to the layer actually in contact with solid cations, appear to become polarised by the presence of strong surface electrical fields (Fripiat and Herbillon 1971). This results in a high degree of dissociation in which the ratio of $H^+ : H_2O$ may be in the order of 1 per cent, equivalent to the acid strength of an N/100 solution. Although there is normally more than one adsorbed layer in natural weathering situations which must lead to dilution, wetting and drying—especially within smaller pores—is likely to lead to acidic solutions. Furthermore, H^+ can combine with water molecules to form an unstable H_30^+ on silicate surfaces. These processes have been expressed by Fripiat and Herbillon (1971, p.16):

$$M^{z}\underset{\text{ads}}{+} H_20 \,|\rightleftarrows|\, M^{(z\text{-}1)}\underset{\text{ads}}{+} O \,|\, + H^+$$

$$H^+ + H_2O \rightleftarrows H_3O^+$$

They lead to a partially irreversible hydration by adsorption of water, followed or accompanied by a first stage in hydrolysis involving a cationic exchange between H^+, H_3O^+, and alkali or alkali earth cations in, for instance, the feldspar lattice. These may be regarded as the first essential steps in the breakdown of silicate minerals into clays.

Groundwater pH may also be affected by the nature of individual rocks: the more 'acidic' types liberating more silica to give an acid reaction (Loughnan 1969). Common pH values for water in widespread tropical environments are cited in table 2.

TABLE 2

COMMON pH VALUES FOR WATER IN DIFFERENT TROPICAL ENVIRONMENTS

Source of water		*Common pH range*	*Extreme values of pH*
Rainfall		5·1–7·4	
Runoff	—sheet flow	4·0–8·5	
	—small streams	4·8–7·6	4·0 (A)–8·8 (B)
	—large rivers	5·0–7·4	
Groundwater	—superficial	5·8–8·6	
(under forest)	—at depth	4·0–8·5	3·3 (C)

Figures taken from several sources including Rougerie (1960). Extreme values from Bleackley (1964), (A); Douglas (1968), (B); Bakker (1960), (C)

Variations of pH also occur with depth. For instance Nye (1955) recorded pH values in a deeply weathered gneiss near Ibadan, Nigeria. He found that an acidic reaction (pH 5·5–6·7) persisted to a depth of at least 5 m, below which there was a sharp rise to a pH value of 8, associated with the water-table.

In general there is an absence of soil environments with acidities exceeding pH 4·5, indicating (figure 5) that alumina and ferric iron are seldom likely to go into solution from this cause. Similarly the absence of alkaline groundwater above pH 9 suggests that alumina will remain poorly soluble throughout the range of soil water environments. On the other hand iron in its reduced form goes readily into solution within the range shown.

Aluminium does, however, become mobile in soils rich in organic matter, where the pH descends below 4. Such is the case with podzols, but suitable conditions occur rarely in the tropics.

Soil temperatures on lowland tropical sites remain very nearly constant at around 28°C below about 50 cm, but fluctuate widely at the surface. D'Hoore (1964) found variations from below 25°C to 39°C at a depth of 19 cm at Yangambi in the Congo, while surface temperatures may exceed 45°C in semi-arid regions. At this temperature level, biological reactions are inhibited, and the soil is generally desiccated. However, in humid tropical conditions, within free draining soils a slightly acid environment at a temperature of around 28°C persists for the greater part of the year.

In detail the weathering environment will exhibit significant departures from mean figures, but it is clear that in the humid tropics hydrolysis is promoted by abundance of moisture and high temperatures. Iron in its reduced form will be very mobile, but will become immobilised by oxidation. Silica solubility appears to be moderate for amorphous forms, but the effects of solution on quartz are probably minimal. Alumina under these conditions is not mobile, but remains in the profile to form clay minerals or as gibbsite.

The role of organic matter and biological processes in weathering

A critical role in weathering processes may be played by both living organisms and by organic compounds resulting from the decay of plant materials. However, the quantitative importance of these factors is as yet difficult to assess. The cycling of nutrients within the vegetation cover is but one aspect of this problem, and it has been suggested that the deep rooting trees of the tropical rain forest are capable of removing important quantities of dissolved mineral matter from considerable depths (Lovering 1959 quotes depths of from 10 to 30 m as not uncommon). As a result of continual or periodic leaf fall, and of the decay of woody material, much of this may be returned to the surface layers of the soil. Lovering (1959) has pointed out that many plants accumulate considerable quantities of major elements such as Si, Al, Ca, and Fe, during growth. On the basis of mineral analyses he concluded that in a tropical forest decaying plant material may average 2·5 per cent silica. Each year such forests contribute to the surface between 10 and 20 tons of dry matter per acre. Using 16 tons as an average, Lovering thought that 0·4 ton of silica would be contributed to each acre per year. Over a long period this process amounts to the removal of important quantities of silica from some depth below the land surface. Illustrating the situation by means of a basalt containing 49 per cent silica, Lovering calculated that silica equivalent to a depth of one foot of basalt would be cycled in this way during a period of 5000 years. This uptake of silica by plants might also partly account for the low silica content of tropical groundwater which lies in the range 8–16 parts per million (p.p.m.) for vadose water and 30–40 p.p.m. for water extracted from deep wells and tunnels (Harrison 1934; Lovering 1959). Davis (1964) gives 17 p.p.m. as a median value for groundwater. These are low figures when compared with the theoretical solubility of silica which at 25°C is around 135 p.p.m. (Krauskopf 1967). Not all the silica returned to the soil as leaf litter and tree fall remains in the surface soil; some is inevitably lost as runoff or throughflow. But within a mature rain forest environment such losses may be comparatively small, except during occasional, very heavy storms.

An important group of processes results from the presence of organic acids in groundwater. These are products of plant decay and are principally humic and fulvic acids. They have a marked effect upon the solubility of certain elements liberated during weathering, particularly iron. Strakhov (1967) for instance claimed that one half of the Fe, Mn, Ni, and Cu found in the Dneiper River is contained in the form of organic complexes. Such organic complexes take the form of ring structures within the centre of which the metal cation is held by N, O, or S atoms (see Krauskopf 1967, Fig. 11.1). They are called *chelates* and the process by which they remove metal cations is often described as chelation or cheluviation (Pedro 1968; Carroll 1970).

Chelation is imperfectly understood and may not in fact be the only mechanism by which iron and other metal cations are taken into solution by organic acids. Some, such as sodium, may form soluble salts such as acetates by substitution for H^+ ions in carboxyls, whilst others such as lead may form metallo–organic compounds by linkage with carbon atoms (Krauskopf, 1967). The removal of iron in the weathering process is of critical importance to the understanding of such widespread phenomena as podsolisation and lateritisation which is so important in the tropics. Under weathering conditions it can be shown that this must involve reduction of the iron to the ferrous form (Fe^{++}), or occur as a result of the intervention of organic compounds. Reducing conditions seldom occur in freely drained sites, where oxygen is present (Eh generally remains above +400 mv).

Several authors (Schatz *et al.* 1957; Schalscha *et al.* 1967; and Ong *et al.* 1970) have attempted to provide a greater understanding of chelation and related processes. Schalscha *et al.* (1967) subjected pulverised granodiorite to reactions with several acids including salicylic acid which forms a red coloured Fe-salicylate complex with iron. The possibility that the iron was removed from a solution in water was discarded, because the solubility of the iron at controlled pH values was too low for this to account for the observed results. The authors therefore concluded that the salicylic acid extracted the iron from the rock by direct reaction which was shown to occur independently of pH. Ong *et al.* (1970) were able to treat several metals with a peat extract under controlled laboratory conditions. In experiments with Cu, Zn, Pb, Al, and Fe^{3+}, and with organic acid concentrations of from 4 to 40 p.p.m. they found that 'except at low pH values the amount of metals held in solution/suspension is always greater when organic acids are present' (p. C-133). The amounts increased as the pH was increased (figure 6A), and this was particularly noticeable for ferric iron, the solubility of which normally declines as pH increases. On the other hand, although the solubility of aluminium was increased it continued to have a very low solubility at or near neutral pH (figure 6B). This

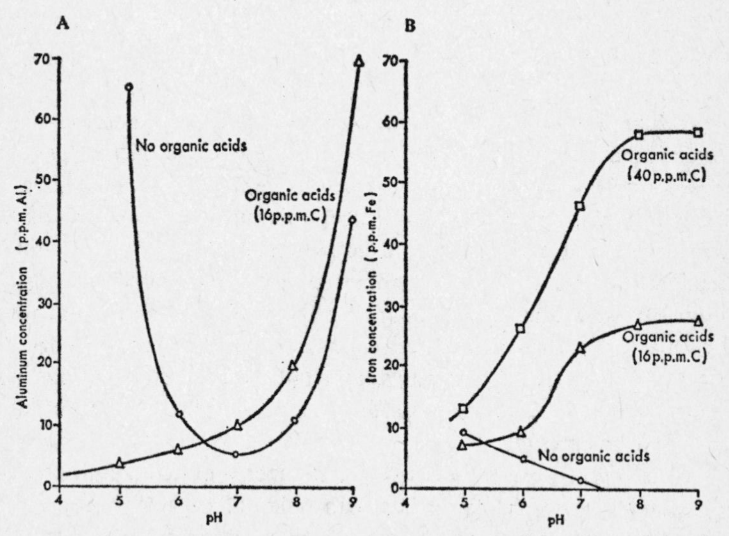

Figure 6 Solubility of aluminium and ferric iron as a function of pH in the presence and absence of organic acids (from Ong, Swanson and Bisque 1970).

A. Solubility of aluminium as a function of pH in the presence and absence of organic acids; B. Solubility of ferric iron as a function of pH in the presence and absence of organic acids.

was held to indicate a mechanism by which iron may be separated from aluminium during weathering under oxidising conditions. Although these authors support the concept of chelation as a major process involved, they think that there is evidence to show that iron forms hydrophobic colloids in the presence of organic acids. Their results showed that, especially in the case of iron, increases in concentration of organic acid and in pH both led to greater solubilisation, and produced comparatively stable compounds providing for transport of the iron in solution.

These results are, of course, significant to a wide range of weathering environments beyond the tropics, but they throw light upon some tropical situations of importance in geomorphology, and particularly upon the development of bauxites and laterites.

Salt weathering

It has been claimed that certain weathering phenomena result primarily from a group of processes known as salt weathering. These arise either from the hydration of rock minerals or from the initial process of crystallisation from supersaturated solutions of adventitious salts (Wellman and Wilson 1965; Evans 1971). The effects of such processes are the result of expansive stresses applied to grains or crystal boundaries and are seen mainly as granular disaggregation or scaling (desquamation). It has been suggested that some weathering pits and taffoni result from salt weathering, especially in polar, arid and coastal areas. The effectiveness of this process within the humid and sub-humid tropics is probably reduced by the humidity of the atmosphere and by the tendency for leaching of dissolved salts. However, since scaling commonly affects the curved rock faces of domed inselbergs, and taffoni are developed within this range of climates, the possibility of a wider role for salt weathering must be envisaged.

It has been shown (Evans 1971) that supersaturated solutions are necessary before crystallisation can take place under common confining pressures in rocks, so that evaporation from rocks and rock surfaces between storms is unlikely to be effective in climates where leaching is active. However, the process of hydration undoubtedly occurs within most weathering environments and may be effective in breaking down rock fabric in advance of important chemical changes. Hydration really involves only the adsorption of water, which in the case of silicate minerals is accompanied by hydrolysis. It is therefore difficult to isolate as a separate weathering process. Hydration is probably primarily responsible for scaling on crystalline rock faces, and it may be important in granular disintegration of granites, especially where the fabric of the rock is weakened by microfissures or other planes of weakness allowing deep penetration of small amounts of water. The process may also be important to taffoni formation in locations where salt crystallisation is unlikely.

Weathering processes and the water-table

Until recently many writers claimed that the processes of alteration virtually ceased under saturated conditions below the water-table. Thus Reiche (1950, p. 84) considered that 'effective weathering extends approximately to the water-table, although groundwater circulation permits some alteration to greater depths in formations or rock zones of high permeability.' This reasoning was followed by Linton in his study of tors (1955). However this contention seems to be unsupported, and in fact early writers such as Campbell (1917) thought that alteration might continue to depths of 30 m below the watertable.

Nye (1955, Part II, p. 61) concluded from studies in western Nigeria that rock had become altered 'considerably below the water-table at 28 feet, mainly by decomposition of feldspars'. He distinguished between a zone of alteration at these levels and a zone of more thorough weathering above the water-table. This concept is also advanced by Lelong and Millot (1966) following de Lapparent (1941). These authors recognise a 'zone d'alteration inferieure' within which slow water penetration via 'microfissures' and cleavage planes in the rock leads to the alteration of plagioclese and other silicate minerals to sericite, and of ferro–magnesian minerals to chlorite. This they distinguish clearly from the 'zone superieure d'alteration', where more rapid circulation of groundwater intensifies the process of hydrolysis with the formation of kaolinite and other clay minerals.

The rate of alteration in the upper vadose zone of alternate wetting and drying is however thought to be very much higher than that at greater depths. Birot (1968) claims that a factor of 20 to 30 times may be involved, and whether or not this can be accepted, there is general agreement on the need for repeated leaching of the soluble products, to promote both the speed and the direction of rock weathering (Keller 1957). Ollier (1969) has also referred to this question, pointing to the misconceptions involved in much earlier work.

The relationships between surface landform, weathering depth and water-table are complex and by no means fully understood. Some morphological aspects of the problem are considered further in a later chapter, but it may be noted that the comment offered by De Swardt and Casey (1963, p. 9) regarding the weathering of sandstones in eastern Nigeria, that 'the bottom of the zone of weathering is thus clearly not related in depth to the present landsurface or to the watertable', is true also of the relationships within many crystalline rocks in the tropics. Watson (1964) for instance commented that granite in Rhodesia was observed to be weathered to well below the level of the water-table, even at the end of the dry season.

A recent and most thorough study of water-table relationships in the tropics has come from Lelong (1966), who studied four experimental wells near Parakou in Dahomey. The climate here is markedly seasonal, 1200 mm of rainfall being concentrated into the months May–October. The wells were sited on a gently undulating surface of erosion beneath which weathering depths in Precambrian gneiss were generally at least 10 metres. Lelong found that the water-table did not act as a true discontinuity, and that from a depth of about 2 m some two-thirds of the pores were already saturated. From an analysis of the filling of larger pores during the wet season he concluded that the weathered gneiss had a 'useful porosity' of only 5–6 per cent. This porosity over a depth of several metres absorbed the early rainfall of the wet season without any marked rise in the water-table. This he called the period of 'occult recharge' and it was followed by a rapid rise of the water-table in July (after about 600 mm of rainfall). At the end of the rains, a rapid descent of the water-table was observed, at a rate of about 1·30–1·50 m per month, until a low water level was attained, after which further lowering was very slow. This period of water loss was interpreted as an emptying of the larger pores, together with some further loss of humidity.

The observations showed that a transference of water from deep levels towards the surface takes place each dry season, under conditions of non-saturation. This movement Lelong interpreted as an isothermic diffusion responding to a gradient of concentration. An important corollary of these findings might be the possibility that the products of hydrolysis formed at depth in the profiles can migrate towards the surface layers by a similar process of ionic diffusion. In this way evacuation of solutes from the base of deep weathering profiles becomes possible, and this may offer a solution to the problem of the formation of deep and discrete basins of weathering in crystalline rocks (see chapter 3 and figure 21). The process of water loss and renewal also shows that renewal of groundwater is not wholly dependent upon actual flow of water at superficial levels, but may penetrate to considerable depths each wet season.

Early stages of weathering penetration and rock alteration

According to Harris and Adams (1966) 'during the early stages of alteration there appears to be a general dissolution and etching along twin planes and grain boundaries.' Such attack in granites tends to affect first the calcic plagioclases, and Ruxton and Berry (1957, p. 1267) observed that 'when part of the plagioclase has decomposed, and the orthoclase is beginning to be attacked, the rock breaks down into platy fragments of decomposed granite called gruss. Plagioclase completes its decomposition first, and when most of the orthoclase has been rotted to kaolin the gruss crumbles into a silty sand.'

Admission of water within the rock on the macro-scale occurs by penetration along joint planes. But Birot (1962) has called attention to the presence of 'microfissures' (or 'microcracks', Bisdom 1967) which are almost certainly minute fractures due to forces resulting from crystallisation, tectonics or relief stresses. In some cases these will approximate to what Chapman (1958) called 'potential joints' represented by 'cracked grains, cleared grains, disjointed grain boundaries, tiny faults and layers of tiny fluid inclusions' (p. 555). Such minute fractures almost certainly form planes of weakness along which weathering penetrates, particularly in the superficial layers of the rock.

The effects of rock texture on weathering penetration are fundamental. In coarse granite for instance, the early oxidation of biotite which has been observed by several writers (including Eggler *et al.* 1969) may lead to the disaggregation of the rock into a gruss, before much fundamental chemical change has occurred. In finer rocks the decay of one mineral generally has less effect. It is thus often asserted that coarse grained rocks weather more rapidly than fine textured rocks. But contrary observations can be cited, particularly in relation to the rapid decay of some volcanic rocks such as basalt. However, since mineralogy influences weathering so greatly, the isolation of texture as a factor in the weathering of crystalline rocks is very difficult.

It is important to add that at this early stage of weathering, disaggregation results from an increase in volume due to hydration; at a later stage when decomposition has proceeded further, this is compensated by removal of solutes by leaching, Dennen and Anderson (1962), and widely varying estimates for total losses of material in solution have been given. Ruxton (1958) gives some figures for granite in the Sudan which suggest losses of around 34 per cent; other calculations suggest much higher figures, but since the bulk density of weathered materials may be less than 1, calculations based on unit volume must be treated with caution (Lovering 1959). At this stage a greatly increased porosity develops, which in turn aids the penetration of groundwater to new levels. Lumb (1965) has shown that in the granites of Hong Kong theoretical and actual figures for clay percentage and porosity varied by an amount suggesting leaching and eluviation of fines of between 10–20 per cent. The effects of such removal on landform development may be important. It has been pointed out that porosity is increased by this means as the voids ratio rises without any necessary shrinkage of the rock—expansion from hydration having previously occurred.

Ollier (1967) has referred to evidence in favour of constant volume alteration during weathering at depth. The detailed preservation of rock textures and structures, from the formation of pseudomorphs of original minerals to the retention of veins, joints and other features of the rock fabric all argue for very little volume change during weathering. Such features are not preserved in the upper levels of profiles, but are generally seen at depth. However, in the later stages of weathering into kaolinite, and particularly into sesquioxide rich residues, loss of volume is known to occur. This is referred to by Haantjens and Bleeker (1970) who calculate a volume decrease of 23–37 per cent for sesquioxide weathering in New Guinea. Loss of volume on weathering is basic to an hypothesis of laterite formation adduced by Trendall (1962) and is discussed in chapter 2.

The progress of weathering in crystalline rocks

Keller (1957) has emphasised the importance of removing the products of hydrolytic weathering if the reactions are to continue. He listed the following means by which ions liberated during weathering may be removed:

(1) repeated leaching by fresh rain;
(2) introduction of H^+ ions which:
  (a) will combine with OH^- ions, thereby removing them as water;
  (b) displace by cation exchange metal cations from the metal compounds;
(3) precipitation of ions as relatively insoluble compounds;
(4) removal of ions by chelators (complexing);
(5) absorption and assimilation of the products by living plants and animals;
(6) absorption of the products by colloidal substances.

Leaching is important to the course of chemical reactions, because it maintains the system as a more or less stable environment, within which water is close to neutral in reaction. It also carries soluble products out of the system altogether. If leaching ceases as a result of dry atmospheric conditions then the concentration of solutes will rise, affecting the pH and the oxidation potential of the weathering environment.

The predominance of kaolinitic clays in association with iron and aluminium sesquioxides, particularly within the humid tropics, is clearly a reflection of this factor. Kaolinite clays are produced under conditions of free drainage and slight acidity of groundwater. Where deep regoliths occur beneath well drained and forested sites such conditions are met with for a large part of the year. The excess of H^+ maintains silica in the system, so long as the divalent ions which flocculate silica are removed. In practice these are Ca^{2+}, Mg^{2+} and ferrous iron (Fe^{2+}). The first two are easily removed in solution where strong leaching occurs, and the unstable ferrous ions are rapidly immobilised by oxidation to the ferric state under oxidising conditions. Although monovalent Na^+ and K^+ may be present they are commonly also leached, but in any case do not flocculate the silica. Alumina is retained in this system, and the insoluble products of such a weathering environment are kaolinitic clays (including halloysite), and gibbsite where leaching is particularly intense, together with the sesquioxides of iron and aluminium, plus of course residual quartz and perhaps mica.

The peculiarity of the tropical environment would appear to be the thoroughness with which this process is carried out. The conversion of all the silicate minerals to clay, and iron and aluminium compounds, mostly within the clay-size fraction, produces a bimodal regolith having a medium–coarse sand fraction dominated by quartz and a clay fraction dominated by kaolinite. Such a regolith has important characteristics resulting from its particle size and mineralogy. The presence of a coarse sandy material and the absence of swelling clays provides a highly permeable regolith which can remain stable under conditions of prolonged, high intensity rainfall, and which also admits groundwater to continue alteration of fresh rock beneath very thick mantles of regolith.

The conditions promoting these reactions are continuous high temperatures, together with the presence of organic matter, and associated with frequent rainfall maintaining the moisture supply within the regolith.

But from earlier comments it is necessary to envisage a phase of partial decomposition of the silicate rocks below the watertable. Evidence suggests that this environment will be alkaline, and it is clear that removal of soluble cations will take place only slowly if at all. Hydrolysis under these conditions will produce different minerals which in terms of later alteration may be regarded as intermediate products. It has been suggested that sericite and chlorite may be examples of such minerals in which cations such as Mg^{2+} are retained in the silicate lattice.

Suggested sequences of alteration for common rock forming minerals have been offered by many writers. A useful summary (table 3) has been offered by Fieldes and Swindale (1954). This indicates the formation of gibbsite as a transitional product in kaolinitic weathering of feldspars, a phenomenon confirmed by later studies (Krauskopf 1959; Sanches Furtado 1968; Eden and Green 1971). However, gibbsite may also be an end product under intense leaching of silica and iron. This is illustrated by Craig and Loughnan (1964) on the basis of studies of basic volcanic rocks in New South

TABLE 3

WEATHERING OF PRIMARY ROCK-FORMING MINERALS

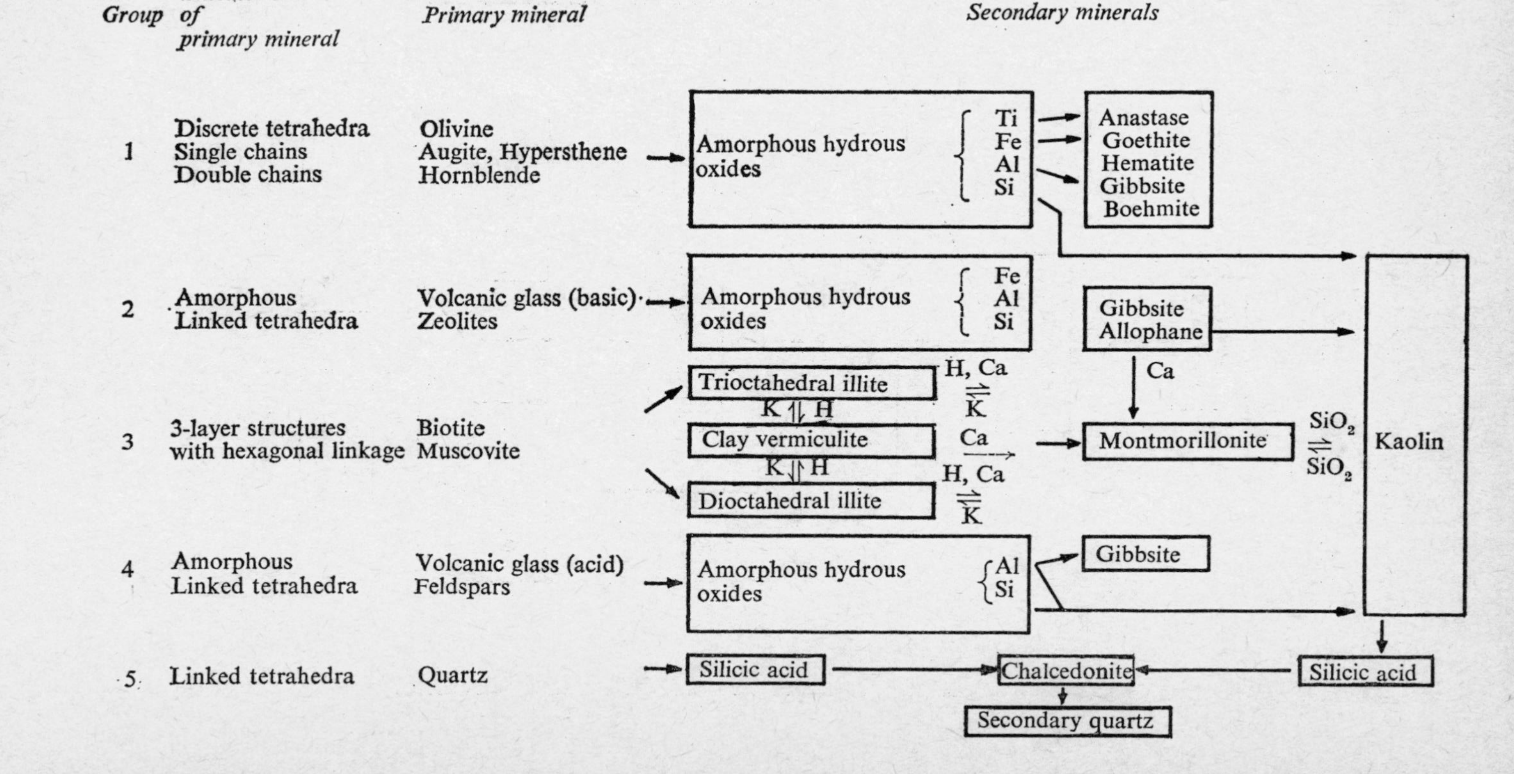

Wales (figure 7). This study also points to the intermediate formation of montmorillonite in the weathering of plagioclase and olivine, to create a sequence of weathered products in response to increased leaching.

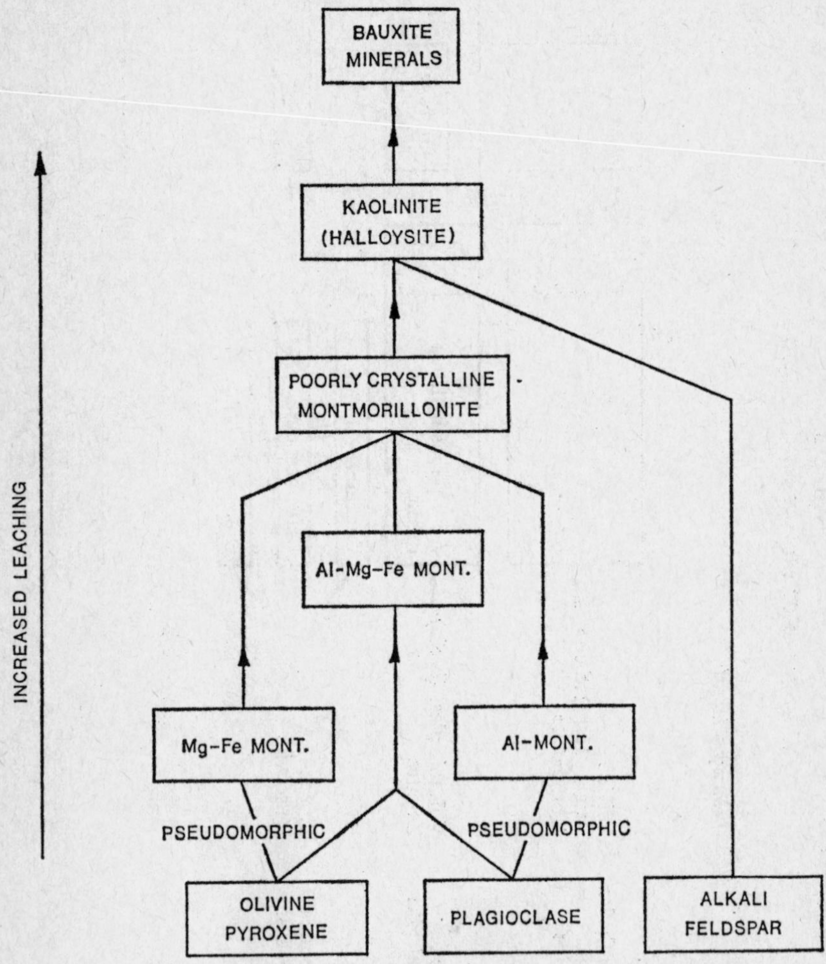

Figure 7 Diagrammatic representation of the weathering sequence developed on Tertiary basalts in New South Wales (from Craig and Loughnan 1964).

Kaolinitic weathering is not the only widespread type in the tropics. Actual mineral products in regoliths will depend in part on the mineralogy of the parent rocks, but also on variations within the weathering system.

Rocks rich in aluminium will weather to form gibbsite in addition to kaolinite within a weakly acid environment, and the two minerals are commonly associated. On the other hand the presence of large amounts of organic matter may reduce the pH to below 5 and aluminium as well as iron may be complexed out of the system. Such an environment does not persist throughout the profile, over wide areas, or continuously for long periods, so that iron and aluminium removed in this fashion may easily be precipitated elsewhere or at a later time. Variations in the weathering environment of a particular horizon may occur with time, with or without the effects of climatic change. Thus if the pH varies from close to the abrasion pH of common minerals (8–11) to the neutral reaction of pure water, minerals such as silica, alumina and iron oxide may become separated, and may be deposited individually as layers within the profile or at locations across a slope catena to form particular deposits such as laterite, silcrete or bauxite. The details of such translocations may be very complex and few generalisations are valid (see chapter 2).

Regional climate and weathering products

In a similar manner, variations in zonal climate affect the weathering environment in important ways. Keller (1957) points out that where potential evapotranspiration (PE) exceeds rainfall, leaching is restricted. Most cations remain in the system and flocculate silica and alumina to form 2 : 1 lattice clays such as illite or montmorillonite. Much of the iron is then incorporated into the clay mineral lattice (Segalen 1971). Such regoliths forming in dry climates will not exhibit lateritic horizons, although saline encrustation may occur. The 2 : 1 clays absorb water readily, swelling as they do so to form a relatively impermeable material subject to much greater runoff and potential erosion hazard.

Between the dry climates and the humid tropics, are the much larger areas of seasonal climate. Within these, leaching and deposition fluctuate across the profiles with the season, and usually in relation to the water-table level. Although kaolinitic clays predominate in many instances, it is particularly within these areas that iron oxide and silica may be preferentially deposited at certain horizons. However, when the topic of laterite is discussed in greater detail it will be seen that seasonal fluctuations in humidity may not be the only, or the most important factor. In nearly all tropical lowland climates rainfall, although it may be frequent, is periodic and drying out of the upper soil horizons may occur between storms.

Attempts have been made to characterise weathering environments by their dominant weathering products, and certain of these have climatic implications. Thus, Pedro (1968) has recently suggested that the tropical zone may be divided into three (figure 2, page 5):

(1) Zone of bisiallitisation with the formation of 2 : 1 lattice (montmorillonite) clays in areas receiving less than 500 mm rainfall.
(2) Zone of monosiallitisation (kaolinisation) in which 1 : 1 kaolinite clays dominate the regolith in areas receiving 500–1200/1500 mm rainfall.
(3) Zone of allitisation in which gibbsite occurs along with kaolinite in areas receiving more than 1500 mm rainfall.

Sanches Furtado (1968) points out that throughout the sub-humid and humid tropics kaolinite dominates most regoliths. But this mineral is accompanied by others where either high humidity or a long dry season occurs. He found that granites within intertropical regions produced weathering residues:

Kaolinite dominant	1000–1200 mm rainfall
Kaolinite plus montmorillonite	800–1000 mm rainfall
Kaolinite plus gibbsite	1200–1500 mm rainfall

Geothite (the principal form of ferric oxide) although present throughout the range was noticeably more plentiful in the samples from humid regions. Segalen (1971) has emphasised that the segregation of iron, often as amorphous ferric oxides, must be an inevitable result of the widespread formation of kaolinitic clays.

Studies of basalt weathering in the Hawaiian Islands (Sherman 1952, quoted by Loughnan 1969) show close agreement with the figures for granite, allowing for the generally greater amount of goethite formed from the ferro-magnesian rich rock (figure 8). This study also illustrates the increasing percentage of clay with higher rainfalls, from 20 to 30 per cent at 750 mm to over 60 per cent at 2500 mm and over. West and Dumbleton (1970) from studies of basalt weathering in Malaysia have demonstrated a much greater

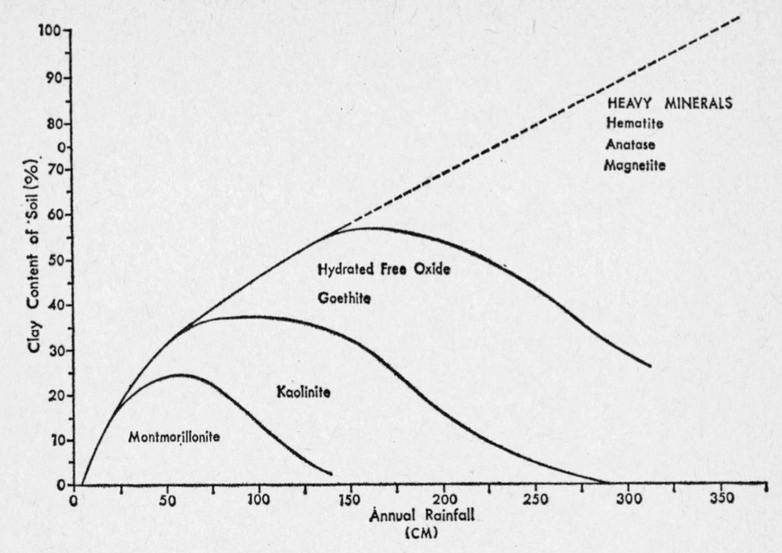

Figure 8 Progressive types of clay development in Hawaiian soils under an alternating wet and dry climate (from Sherman 1952)

abundance of gibbsite in regoliths subject to heavy leaching, a marked increase in this mineral appearing as the rainfall rises from 2000 to 3500 mm per annum.

Weathering products are also influenced by temperature and Jenny (1941) has demonstrated the increase of clay in soils with annual temperature (figure 9). Within the tropics this factor only becomes important at higher altitudes, but because tropical regoliths are normally characterised by high clay content this enables comparisons to be made with weathering products from non-tropical areas.

Bakker, for instance, together with his colleagues at Amsterdam (Bakker 1960, 1967; Bakker and Levelt 1964) have claimed that regoliths from granitic rocks within tropical forest climates will contain 30–50 per cent clay (particle size less than 2 μm) of which nearly all is kaolinite or halloysite, though gibbsite may reach 30 per cent in some cases. These figures correspond well with others given here. Bakker compared these with granite regoliths from central Europe in an attempt to define the weathering environments of this

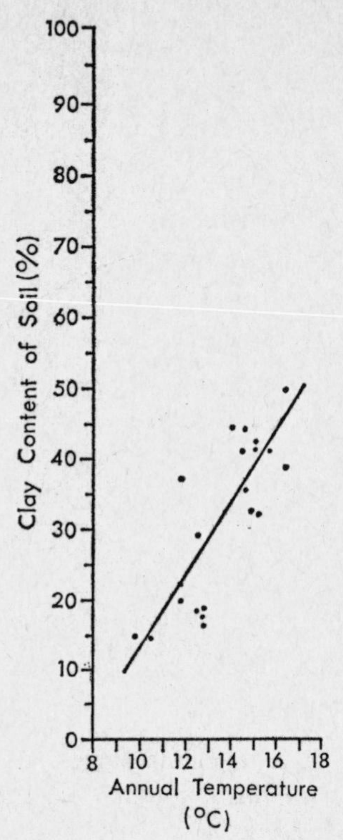

Figure 9 The clay content of soils derived from basic rocks in relation to mean annual temperature (from Jenny 1941).

area during the Tertiary. Some of his conclusions are discussed later (chapter 10).

Degree of weathering and properties of regoliths

The inference from the above reasoning is that a free drainage tropical environment will produce a clay-rich regolith containing residual quartz. It is recognised that this material is produced via poorly defined stages from original rocks of diverse character. A different approach to the question of weathering is to use some index of the degree of alteration. In general there are three methods used to assess this problem:

(1) For exposed rock, authors such as Melton (1965) and Ollier (1965) have suggested a qualitative scale of friability.
(2) For disaggregated rock, usually brought up in cores or sampled from cuttings, particle size distributions together with measures of physical mobility based upon mechanical properties of the regolith have been used by Lumb (1962, 1965) and Ruddock (1967).
(3) For all materials it is possible to devise measures of the degree of alteration on a chemical or mineralogical basis, by comparing a given sample with the composition of the original rock, or by using other indices of alteration. Authors such as Harrison (1934) Reiche (1943) and more recently Ruxton (1968a) and Grant (1969) have considered this problem.

These methods converge with attempts to describe deep weathering profiles in terms of the degrees of weathering. But field descriptions are usually based on morphological criteria, comparable with descriptions of soil profiles, and it is not always possible to relate these directly with the assessments of weathering derived from the methods discussed here.

Knill and Jones (1965), Melton (1965) and Ollier (1969) suggest scales of friability similar to that given below:

(1) Fresh rock: no visible alteration, a hammer bounces from surface.
(2) Stained or pitted rock: no interior alteration.
(3) Slightly altered rock: deep pitting and interior staining; may break easily with a hammer.
(4) Partially decomposed throughout: still cohesive but partially granular; easily broken with hammer or by kicking.
(5) Thoroughly decomposed: readily broken by hand and may disintegrate in water.

Clearly this sequence is relevant mainly to material still sufficiently cohesive to be regarded as rock rather than regolith (except grade 5). No measure of the degree of mineralogical or chemical alteration is offered for these categories and they are therefore difficult to fit into the other schemes. Nevertheless if all these measures of weathering are reviewed together, they illustrate the course of decomposition quite well. Measurement of particle size and mechanical properties in regoliths commonly requires some mineralogical information in addition, and is restricted in the main to the products of the later stages of weathering leading to complete breakdown of the original rock. However, in geomorphology such measurements have great meaning, for they directly affect the physical mobility of the regolith. Their applicability is also wider than might first be thought, because in many deep weathering profiles there is a sharp transition from little altered to disaggregated rock at the basal surface of weathering or weathering front. However, following upon the previous sections some discussion of the chemical method is appropriate first.

Reiche (1943, 1950) formulated two indices of weathering: the Weathering Potential Index (WPI) which is the molepercentage ratio of the alkalis and alkaline earths to the total moles present,* and the Weathering Product Index (PI) which is the ratio of silica to silica and sesquioxides.† As the WPI decreases so mobile cations decrease with increase in hydroxyl water, while on the other hand as the PI decreases so silica content decreases. Both Reiche (1943) and Ruxton (1968a) show how, when plotted against each other, these indices express the progress of weathering. Short (1961) suggested that the WPI be scaled from 0–100 to provide a relative Weathering Index (WI). This is used by Ruxton, who found that the most valuable relationships were obtained by plotting total element loss against the Weathering Index, and the WI against the silica: aluminium mole ratio. He also found that the silica: aluminium mole ratio and the silica loss alone were both significantly correlated with total element loss, and he concluded that the ratio of silica to alumina (or Al plus other sesquioxides) is a good simple expression of the degree of weathering. It is found that this Si: Al mole ratio varies from 5–10 in fresh igneous and metamorphic rocks to 1·5–2·0 in kaolinised weathering residues.

A simple approach to this problem has been advanced by Grant (1969), who shows that if the abrasion pH is found for fresh and altered rock samples, then the pH values decline as weathering

*WPI $$\frac{100\,(K_2O + Na_2O + CaO + MgO - H_2O+)}{SiO_2 + Al_2O_3 + Fe_2O_3 + T_1O_2 + FeO + CaO + MgO + Na_2O + K_2O}$$

†PI $$\frac{100\,SiO_2}{SiO_2 + T_1O_2 + Fe_2O_3 + FeO + Al_2O_3}$$

increases. Rock is simply crushed and ground in a small quantity of water and the pH determined for the resultant slurry. Grant worked on samples of granite and kaolinised regolith from the same rock, and he concludes that the decline in pH is to be attributed to the rapid loss of sodium and potassium and to the corresponding increase in kaolinite. A comparison is made with the Weathering Product Index of Reiche, and Grant suggests that the Abrasion pH is a function of this ratio, and therefore of similar significance.

It is unfortunate that the increase of clay minerals during the weathering process cannot be accurately determined from mechanical analysis. It can readily be shown that clay minerals persist in the silt and sand size fractions of regoliths, whilst non-clay minerals, especially sesquioxides, appear in the clay-size fraction (less than 2 μm). Flach *et al.* (1968) show that an increase in clay less than 2 μm from 25 per cent at a depth of 610 cm to nearly 83 per cent at 25 cm in a weathered andesite from Puerto Rico is due mainly to physical breakdown and increased dispersibility of the clay, and that while 61 per cent of the kaolinite present at 610 cm is in the silt fraction, only 11 per cent remains in this size fraction at a depth of 25 centimetres.

Confirmation of this relationship is found in a detailed study of granite weathering profiles from Ghana (Ruddock 1967). Ruddock's figures (table 4) indicate the expected decrease of feldspar towards the surface, but more particularly show that a high degree of kaolinisation had occurred at a depth of 30 m, the clay mineral appearing over a wide range of particle size. Increase in kaolinite percentage towards the surface takes place mainly within the finer silt and clay grades. The progressively finer texture of the regolith towards the surface at this site is shown in figure 10.

In spite of these difficulties, the use of particle size or grading parameters to indicate degree of weathering is widespread (Bakker 1967; Lumb 1962). Lumb showed that grading properties varied significantly with his index of weathering (Xd).* This can vary from 0·1 to 1·0 and the average for Hong Kong was found to be 0·49, indicating that not all the feldspar was decomposed. He found that there was a 'progressive decrease in average size with concurrent increase in spread and skewness with increasing decomposition'.

*Xd = $(Nq - Nq_0)/(1 - Nq)$, where Nq is the weight ratio of quartz to quartz + feldspar in the soil, and Nq_0 is the same ratio in the fresh rock.

Lumb also found that the voids ratio showed a tendency to increase towards the surface, and with increasing decomposition of the rock (figure 10). But this simple relationship is disturbed by the compaction of the surface soil, due to the seepage forces of groundwater in the soil (Lumb 1962).

TABLE 4

MINERAL COMPOSITION AND PARTICLE SIZE OF SOIL SAMPLES OVER GRANITE NEAR KUMASI, GHANA

Depth of Sample ft.	*Particles less than* 2μm %	*Particles over* 40 μm				*Particles less than* 40 μm
		Quartz %	*Feldspar* %	*Mica* %	*Kao-linite* %	*Kaolinite* %
91	6	19	25	12	11	33
76	5	17	19	7	20	37
46	7	34	12	18	8	28
26	6	26	5	11	11	49
16	17	16	0	8	8	57

(After Ruddock 1967; reproduced by permission of the Council, Institute of Civil Enginners)

Coarse granite soils were found by Lumb to have a low cohesion (0·1 kg/cm^2 where saturation, Sr = 0·5), sufficient to support a 6 m cutting. Finer soils had a higher cohesion which was rapidly reduced to nil by increasing saturation (2·8 kg/cm^2 at Sr = 0·4; nil at Sr = 0·95). In the medium soils permeability was found to be high (circa 3×10^{-5} cm/s^{-1}), and Lumb considered that the surface oxidised zone of the regolith could be regarded as a free draining sand. The high permeability of tropical regoliths derived from granitic rocks reflects the bimodal characteristics of the material and its high voids ratio (0·6–0·9 within 4 m of the surface). Ruddock's study (1967) also suggested a rapid advance of weathering above the watertable; a significant break in the grading curves occurring at this level.

One problem with many of these results is that they apply particularly if not exclusively to granite weathering. Ruddock (1967) for instance found that few systematic variations with depth could be detected within phyllite regoliths from the same area in Ghana, although the proportion of fines increased towards the surface (above 7 m).

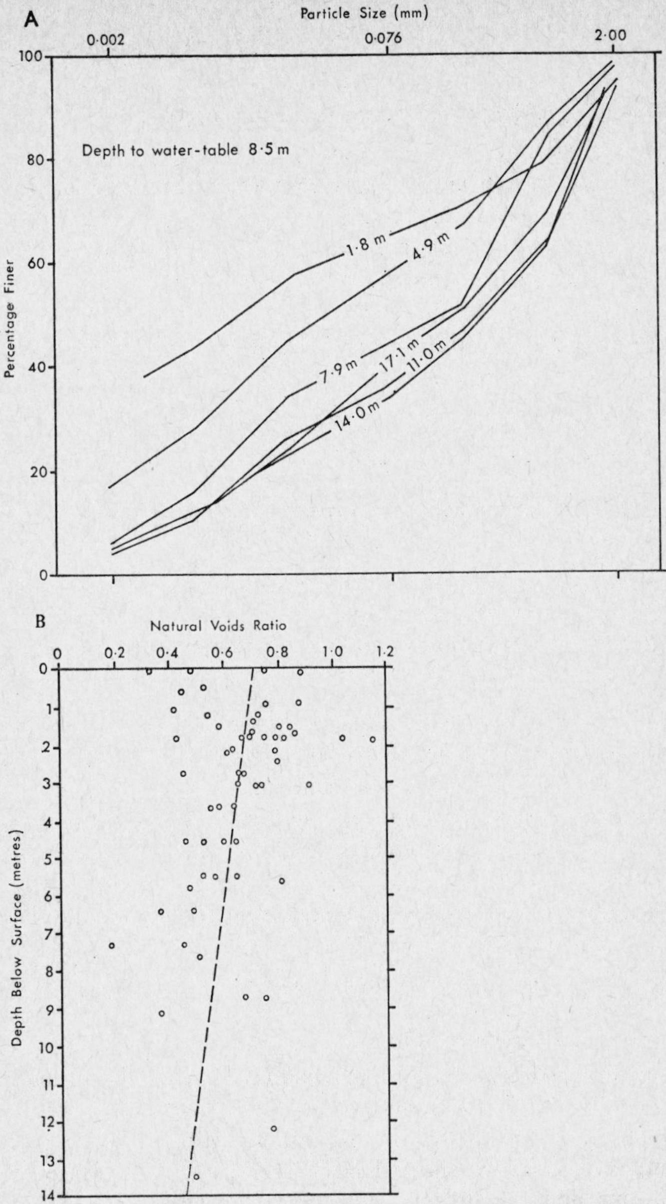

Figure 10 Variations in grading and voids ratio with depths in decomposed granites (after Lumb 1962, and Ruddock 1967).

A. Grading curves for decomposed granite, showing variation with depth at a site on the University Campus, Kumasi, Ghana (after Ruddock 1967); B. Variation of voids ratio with depth in decomposed granites from Hong Kong (after Lumb 1962).

Degree of weathering and the age of the landsurface

A similar approach to the grading of soil parent materials was adopted by Wambeke (1962), who found that the ratio of silt (2–20 μm) to clay varied significantly with the age of regoliths as determined from the inferred age of geomorphic surfaces. The silt content was taken to represent the reserve of weatherable minerals; the clay content the quantity of 'new' minerals resulting from weathering. The figures advanced for well drained soils with no textural B horizon are:

TABLE 5

SILT : CLAY RATIOS FOR SOILS OF DIFFERENT AGES IN EAST AFRICA

Age of surface	*Granite*	*Schist*	*Basalt*
Recent	0·39	0·96	1·03
Pleistocene	0·27	0·18	0·32
Ante-Pleistocene (probably Tertiary)	0·11	0·11	0·09

(After Wambeke 1962; reproduced by permission of the Clarendon Press, Oxford)

These observations show that as regoliths age, the stage of weathering advances both in terms of further mineral breakdown and the formation of progressively greater amounts of clay and sesquioxides, and in terms of the comminution of residual products to form regoliths containing large amounts of clay and residual quartz which persists mainly in the sand fraction. Thus, highly weathered materials from granitic rocks tend to be bimodal in size characteristics, and have a very low silt content. This progression of weathering can be seen within a single profile, as well as by comparison of profiles of contrasting age.

Most of these comments have been applied to tropical weathering on old landsurfaces, but as Wambeke (1962) has shown, marked differences occur between surfaces of varying ages. This becomes very important to studies in tropical geomorphology, whenever attempts to compare shield landsurfaces from, say, Africa with landforms and regoliths from tectonically active zones such as New Guinea. Haantjens and Bleeker (1970) have recently analysed weathering residues from New Guinea, and find that 'maturely' weathered regoliths of the kind referred to above are comparatively rare, in spite of high temperatures and rainfalls in excess of 2000 millimetres. This is clearly due to the unstable relief forms and the rate at which the altered mantle is removed across steep slopes, often in excess of 30°. Most of the regoliths were described as 'skeletal' or 'immature'. In the former, physical breakdown of the rock is more evident than advanced chemical alteration; in immature profiles significant quantities of clay appear, mixed with slightly altered and broken rock.

The authors recognise the fundamental differences between the formation of montmorillonitic clays (which they call 'smectite weathering'), kaolinitic clays ('kandite weathering') and hydrated sesquioxides ('sesquox weathering'). Characteristically, weathering in steeply undulating terrain was found to vary between 3 and 10 m in depth, with very few outcrops appearing even on the steepest slopes. Most of this is kaolinitic, although immature in terms of total rock breakdown. Both goethite and gibbsite (sesquioxides) appear in small amounts, but are rarely dominant, although gibbsite is found in larger amounts on some freely drained sites in very high rainfall areas (over 3500 mm). Lateritisation, a particularly advanced form of weathering considered separately (chapter 2), clearly does not occur under the conditions described. Mature weathering in fact is found in New Guinea mainly on Pleistocene piedmont plains and benches, on a few pre-Pliocene summit plateaux, and on volcanic agglomerates such as are found on the Sogeri Plateau, where 32 m of weathering have been described.

West and Dumbleton (1970) also give comparative figures for granite regoliths in Malaysia, formed under different slope conditions, but receiving the same rainfall (*c.* 2000 mm). On a 23° slope the granite regolith exhibited 37 per cent clay at the surface, but contained significant amounts of illite as well as kaolinite, along with important amounts of residual feldspar and quartz. On the other hand beneath a 3° slope the surface clay content rose to nearly 50 per cent, and was dominated by kaolinite and quartz with some hemetite. Almost all of the feldspar had been decomposed, although contrary to the prediction above no traces of gibbsite were found. Since the formation of gibbsite requires very free leaching of other cations, especially silica and iron, site conditions as well as rainfall are of very great significance to its formation.

Ideas about the time taken for weathering profiles to develop to different degrees of alteration in varying environments remain speculative. Leneuf and Aubert (1960) calculated that complete ferrallitisation of 1 m of granite in the Ivory Coast had taken from 22 000–77 000 years. Ruxton (1968a) found that volcanic dacite glass would alter almost completely to allophane in periods ranging from 8000–27 000 years, but crystalline materials were little affected in this time. Comparing these figures with their own observations from New Guinea, Haantjens and Bleeker (1970) conclude that skeletal weathering predominates in regoliths for up to 5000 years; immature weathering from 5000–20 000 years, and mature weathering in excess of 20 000 years. This corresponds with the findings from the Ivory Coast, but such figures should be treated with caution. Also they give little evidence of time taken to produce weathering profiles of a given thickness, which is important to many geomorphological arguments. In the discussion of laterites Trendall (1962) advances figures that suggest weathering penetration in Uganda may have advanced at a rate of about 9 m per million years, beneath an old lateritised surface. Many of the immature and skeletal profiles in New Guinea will have been produced in a few thousand years, and it is mistaken to equate such widely divergent environments and regoliths.

2 Laterite

The role of indurated laterites in the formation and evolution of tropical relief is so fundamental that it merits extended and careful discussion. The literature abounds with definitions of laterite, but Maignien (1966) has recently reviewed this question and concluded that we must separately distinguish between *indurated laterites* which form crusts of cuirasses when exposed at the surface, often described as *duricrusts* (Woolnough 1927), *plinthite* (USDA 1960) or, *lateritic ironstone* (Pullan 1967), *ferricrete* (Goudie 1972), more loosely defined laterites and lateritic soils which, according to definition, may or may not be regarded as formations which harden on exposure. For geomorphological purposes, the more general term *laterite* will be reserved for 'a highly weathered material rich in secondary oxides of iron, aluminium, or both. It is nearly void of bases and primary silicates, but it may contain large amounts of quartz and kaolinite. It is either hard or capable of hardening on exposure to wetting and drying' (Alexander and Cady 1962).

Indurated laterite may be defined following Du Preez (1949) as 'a mass that may be vesicular, or concretionary, or vermicular, or pisolitic, or more or less massive, consisting essentially of iron oxide with or without clastic quartz, and containing small amounts of aluminium and manganese. Although its hardness varies it can usually be broken and shaped readily with a hammer.'

Although the geomorphological role of laterite is generally limited to the effects of indurated occurrences upon slope forms and evolution, the recognition that these occurrences may develop from non-indurated parent formations which may be much more extensive than the outcrop of exposed duricrust is important to arguments concerning their significance in landform development.

It is generally agreed that induration occurs after exposure of lateritic horizons as a result of erosional stripping of topsoil. This stripping may be confined to breaks of slope along valley sides, but may commonly occur over wider areas, leading to the formation of laterite tablelands, or mesas. The degree of induration is directly related to hardness, which varies with the iron content and crystallinity, and, inversely, with the degree of hydration. The texture of the laterite, and particularly its compactness, also affects its hardness. It appears also that the older, and longer exposed laterites are generally harder.

Aluminous laterites (bauxites) have different formational histories, and are generally less hard, or less well indurated, than iron-rich occurrences. The more or less feruginous deposits which are generally described as laterites grade into aluminous, silicious and even calcareous materials according to the weathering environment. This, like many other weathering problems, can be viewed in terms of the macroclimatic control, in particular the humidity of climate, but local edaphic or site factors may override the zonal pattern. It has been shown here that within the weathering systems elements may be ordered in terms of mobility, and this sequence can be related to the dominant minerals found in duricrusts. In the most humid and freely leached deposits aluminium attains a relatively high proportion due to removal of iron and silica. With progressive lessening of precipitation and reduction of leaching, bauxites give way to laterites, silcretes and calcretes in turn, according to the sequence: Al—Fe—Si—Ca—Na (most mobile).

Of course most duricrusts contain more than one of the oxides indicated above and strict classification on mineralogical or chemical grounds is difficult. Nevertheless, Dury (1969) has suggested that duricrusts in Australia may be classified on a Ternary diagram (figure 11); and described as allitic, ferrallitic, fersiallitic, or siallitic according to the chemical content. The application of such a scheme may be difficult because the chemical content of duricrusts can vary quite markedly over short distances, in response to both variations in rock mineralogy and to site factors.

In this study the more generally used terms will be retained. Crusts rich in aluminium are described as bauxites; those with a high silica content as silcretes, whilst the wide range of intermediate compositions fall into the category of laterites. In a subsequent section a discussion of the transition, particularly between laterites and silcretes, will enquire further into this problem.

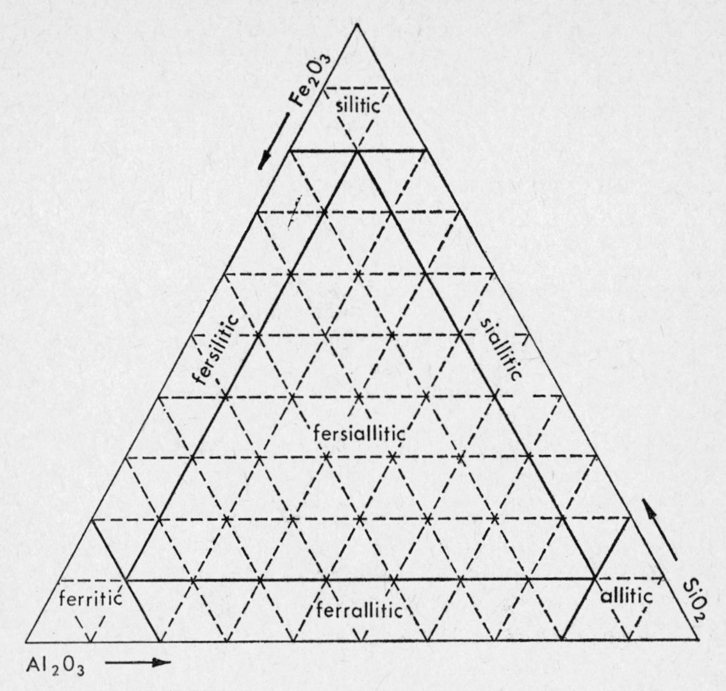

Figure 11 Ternary diagram suggesting boundaries among seven types of duricrust in the fersiallitic range (after Dury 1969)

Laterite structure and morphology

Important differences exist in the morphology of different laterite occurrences, and these are significant both to our understanding of the origins of the laterite and to the discussion of landform evolution in laterite terrain. Three basic structures in laterites are shown as A, B and C in table 6 which follows work by Maignien (1966) and Pullan (1967).

This classification corresponds to some degree with the widespread distinction made by many authors between *primary* laterites of types A or B and *secondary* laterites of type C (Du Preez 1949; Pullan 1967). Primary laterites are regarded as having formed *in situ* from underlying parent materials, but with the possible (even probable) addition of iron from laterally moving groundwater. Secondary laterites on the other hand are seen to have been derived

TABLE 6

CLASSIFICATION OF LATERITES BY MORPHOLOGY

A. Indurated elements form continuous skeleton:
   Vermicular — having small sinuous worm-like tunnels
   Vesicular — having small bladder-like cavities
   Cellular — having many small cell-like cavities, possibly connected.

B. Indurated elements form free concretions or nodules in earthy matrix:
   Nodular — concretions lack concentric lamination
   Oolitic or Pisolitic — concretions exhibit concentric lamination.

C. Indurated elements cement pre-existing materials
   (which may be indurated fragments of older laterites)
   The general terms 'concretionary' or 'detrital' are often used for these laterites but they may be divided:
   Recemented conglomeratic — containing rounded fragments of A types
   Recemented breccia — containing angular fragments of A types
   Recemented nodular — containing fragments of B types.

from the primary deposits as a result of denudation and the recementation of laterite fragments. Primary laterites have been associated with summit features; secondary formations with benches along valley sides or with other low level sites.

Unfortunately this simple distinction is subject to a number of objections. The occurrence of detrital material in well developed laterite sheets forming plateaux and mesas indicates that not all these materials are free from transported fragments (Lamotte and Rougerie 1962). On the other hand it does not appear justifiable to insist that all lower slope laterites are a result of the recementation of older duricrust fragments. In the case of nodular deposits it is not always easy to distinguish *in situ* from recemented occurrences, a problem noted by De Swardt and Trendall (1969) in Uganda. This type of distinction is associated too with the notion that primary lateritisation is associated with one major period, commonly thought to be during the middle Tertiary, when laterites formed as more or less continuous sheets over wide planation surfaces.

It is not the intention to refute this suggestion entirely, but as a weathering process, lateritisation will be continuous so long as suitable conditions prevail. Thus throughout wide areas of the tropics, laterite formation may have been uninterrupted over long periods of geological time. Destruction of older duricrusts has led to the incorporation of fragmentary material in developing laterites which may also be found in some of the oldest surviving deposits, suggesting a still earlier laterite cover. For these reasons the association of laterites of particular morphology or texture with specific formative conditions must be approached with great caution. The concept of a 'primary' laterite, like that of an 'initial' landsurface, imposes an arbitrary and quite artificial limitation to the study of origins and development of existing forms and deposits (Sombroek 1971).

The laterite profile

There are many references in the literature to a characteristic laterite profile, but it is also equally clear that this profile is not ubiquitous and that many variations can occur. Prescott and Pendleton (1952) quote the German writer Walther as the first to characterise the laterite profile in the following terms. Below the ferruginous duricrust he described a *mottled zone,* beneath which occurred a bleached, sometimes almost white, *pallid zone* leading down into parent rock. A complete laterite profile might be described thus. (see also figure 12, and plate 1):

TABLE 7

THE IDEAL LATERITE PROFILE

Zone	*Thickness*	*Characteristics*
Surface soil	0–2 m (exceptionally to 4 m)	Usually grey-brown, slightly humic, and leached topsoil containing many indurated ferruginous concretions or fragments. This horizon is commonly a product of surface wash or mass movement, and may not be directly related to underlying horizons.
Indurated horizon (Crust)	1–10 m (local thickness of 25 m known)	Indurated horizon, or crust, usually red or ocherous in colour but variable in detailed morphology (see table 6). If covered by topsoil the deposit is unlikely to be highly indurated.
Mottled clay	1–10 m	Mottled clay horizon of very variable thickness, often with abundant quartz grains; may have polyhedral or finely prismatic structure.

Zone	*Thickness*	*Characteristics*
Pallid zone (Lithomarge)	up to 60 m (generally less than 25 m)	Pallid zone of kaolinitic clay and quartz sand. Described particularly from Australia, this zone is often poorly developed.
Transitional zone	up to 60 m	Transitional zone of partially weathered rock, containing remnants of rock structures. It is probably uncommon for a thick transitional zone to coexist with a deep pallid zone.
Parent rock		

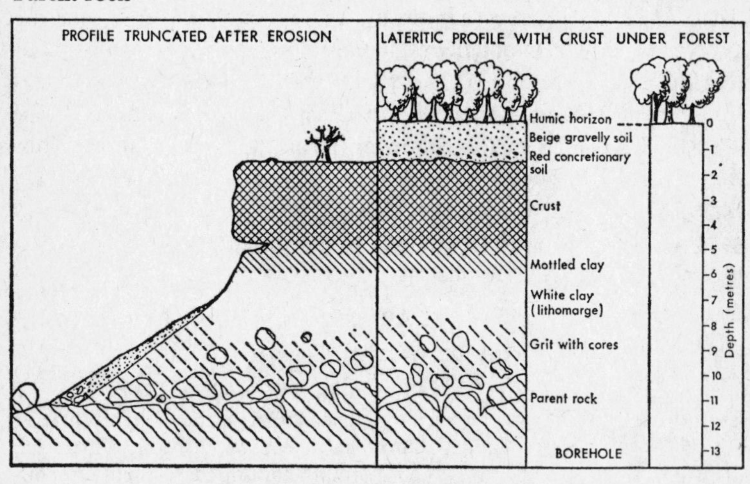

Figure 12 Lateritic profile with ironpan under shade forest, and its equivalent truncated by erosion in the Sudanese region (from Millot 1970).

Departures from this so-called 'classic' profile arise in a variety of forms. The most common variation is an absence of topsoil which may be presumed to have been removed by erosional agencies following the formation of the lateritic horizon. In this discussion of laterites it is axiomatic that the indurated horizon should be present, but this obscures the fact that in some cases a mottled clay horizon is found, having a cover of topsoil, but without any indurated zone. In western Nigeria, Moss (1965) has referred to such mottled clay horizons as 'soft plinthite', and there is a fairly general assumption that induration of the mottled clay leads to crust formation. In fact occurrences of hard laterite are nearly always associated with a mottled clay horizon, and may not always be easily distinguished from it.

It is the pallid zone which probably gives rise to the most widespread disparity between different accounts of laterite profiles. In south-western Australia, Mulcahy (1960) claims that this horizon varies in thickness with the amount of rainfall from 20–25 m in more humid areas to 3 m where the rainfall is 330 mm. But Jessup (1960) has described depths of from 20–60 m in the south-eastern Australian arid zone which indicate that past climatic regimes have been responsible for the surviving depths of the pallid zone. On the other hand Maignien (1966) describes in detail a number of profiles from West Africa in which the pallid zone is not developed as a distinct feature. However, although the pallid zone is not universally present, there is no doubt that it is found widely.

The distinction between pallid zone and the transitional zone of weathered rock is often difficult to make, particularly where the former has not been leached to the same extent as the type examples. In many instances it is likely that, below the mottled zone, further zonation of the weathering profile is more usefully described in terms of degrees of alteration.

The occurrence of indurated laterites

Laterite becomes most important to the geomorphologist when it is exposed through the removal of topsoil. It then becomes hardened or indurated and leads to the development of distinctive landforms. Occurrences of exposed laterite are found in a variety of situations:

(1) As tabular horizontal or sub-horizontal exposed surfaces usually forming hilltops.
(2) As undulating upper surfaces, commonly associated with the occurrence of enclosed depressions.
(3) As exposed ledges forming breaks of slope around upper hillslopes but not exposed on summits.
(4) As bench-forming features on valley sides.
(5) As recemented footslope accumulations or laterite pavements.

A division into summit laterites and lower slope laterites has been suggested to correspond broadly with distinctions between primary and secondary deposits. Thus, Du Preez (1949) referred to *higher-*

lying or peneplain laterite (types 1–3 here, and A and B in table 6) and to *lower-lying or pediplain laterite,* and this designation corresponds with De Swardt's recognition (1964) of a widely occurring *Older laterite* and a *Younger laterite* in tropical Africa. Such simple twofold divisions tend to ignore the complexity of laterite occurrences in many field areas (Sombroek 1971), and may be misleading in terms of genesis. It is clear that indurated laterites occur on summits, intermediate slopes, lower slopes and in depressions. Many of those not on summits appear to have developed in zones of convergent sub-surface water flow (see figure 13, p. 64).

The distribution of laterites

In spite of increasing knowledge it is not possible to present a clear picture of the global distribution of laterite formations. Maps presented in 1952 by Prescott and Pendleton have been superseded for certain areas such as Africa (see D'Hoore 1964), and Australia (Aitcheson and Grant 1970, Stephens 1971). But the picture remains confused, because formations of different kinds and origins are involved. In the first place lateritisation as a process appears to require conditions different from those which lead to exposure and induration of laterite formations. Two different distributions are therefore involved. Secondly Maignien (1966) points out that non-lateritic ferruginous incrustations have also been, mistakenly, equated with laterites. These usually form in semi-arid regions. The same author regards 1200 mm as a lower limit of precipitation for effective lateritisation in acidic rocks. A lower figure of about 1000 mm may operate for basic rocks. There seems to be no upper limit of rainfall inhibiting lateritisation which is known to occur in Guinea and Sierra Leone, where the precipitation exceeds 3000 mm. In this view the lateritisation process is associated with forest or woodland environments, and although periodic drying out of the soil to permit the immobilisation of iron within the lateritic layer is necessary, laterites are found in areas such as Malaysia where there is minimum seasonality.

Exposed duricrusts, on the other hand, are characteristic of the open savanna environments which have a lower rainfall and distinct dry and wet seasons. Within the forest, the organic matter content of the forest soil, together with the pronounced humidity of the surface soil layers are both factors leading to the organic complexing and mobilisation of iron near the surface. This, together with the low rate of surface denudation under the forest canopy, possibly permits the lateritic horizon to migrate downwards as it develops. Degradation of the forest, or climatic change leading to more open vegetation cover seem likely to promote immobilisation of iron and accelerated removal of topsoil, leading to the exposure of the laterite as an indurated duricrust.

Such indurated formations are generally absent from the inner core of both desert and equatorial rain forest. But many 'fossil' laterites are today found in areas which are both climatically and morphologically unlikely to promote contemporary lateritisation. Many of these formations are at least Pliocene in age and some may date from the early Tertiary. All may be taken to represent a prolonged period of weathering and pedogenesis. In many instances, therefore, changes in climate may be held to account for anomalies of distribution. It is also clear that laterite development is sensitive to local conditions of soil drainage which control the reduction–oxidation reactions leading to the concentration of sesquioxides in a lateritic horizon. In general, it is agreed that these conditions are confined to surfaces of moderate to low relief (perhaps with local relief no more than 80 m and slopes under 10°). Many fossil duricrusts survive in elevated locations within dissected terrain. The laterites formed on remnants of the mid-Tertiary 'African' surface over Table Mountain Sandstone in Natal (Maud 1965, 1968) are striking examples.

Notwithstanding the difficulties and anomalies, lateritic duricrusts abound within the savanna plainlands of the tropics, and persist into the desert margins, where they merge with silicious or calcareous formations, and into the higher rainfall zones where they appear to become invaded by the forest, or may merge with bauxitic deposits.

The development of laterite formations

For a full discussion of laterite genesis the reader is directed to the appropriate literature (especially: Prescott and Pendleton 1952; Alexander and Cady 1962; Sivarojasingham *et al.* 1962; and Maignien 1966). The present discussion will concentrate upon aspects of laterite genesis which have a bearing on the geomorphological development of tropical terrains. This hinges upon an assessment of the conditions favouring both the mobilisation and precipitation of iron in the soil profiles, and upon an examination of the sources of iron found in laterites, particularly in those of great thickness.

The behaviour of iron in the weathering system is critical to arguments concerning laterite formation, but is made complex by the difference in mobility between the two valency states of iron: the reduced, or ferrous, (Fe^{2+}) state and the oxidised, or ferric, (Fe^{3+}) state. In the oxidised form iron is virtually insoluble throughout most weathering environments (figure 5, p. 21), but in the reduced state iron becomes soluble wherever the pH is less than 8·5. The mobility of iron is therefore conditioned by the presence of a reducing environment, and this is maintained particularly by organic acids which operate partly by chelation, as previously described. But it is also influenced by the availability of atmospheric oxygen, and where this is largely excluded, below the watertable, reducing conditions commonly prevail, unless the groundwater becomes alkaline (Nye 1955). The redox potential (Eh) of the weathering environment is important to this reaction, but although the laboratory conditions are well established (Loughnan 1969), knowledge of Eh under natural conditions is scanty.

However, from these principles it can be seen that iron is likely to become mobilised below the water-table and on sites where organic matter is not too rapidly destroyed, and that it will be precipitated where oxidation immobilises the iron in the ferric state. It is commonly considered that the most common site for immobilisation of iron will be in the zone of water-table fluctuation. The routes by which iron may reach such precipitation zones will now be considered.

It is particularly important when relating the theory of iron mobility to the field occurrences of laterite to consider two distinct but related problems. The first is the need to explain very great local thicknesses of iron-rich laterite deposits, sometimes exceeding 15 m; while the second concerns the conditions of topography that favour the development of extensive laterites.

Laterite formations are widely associated with terrains of low relief. Many early accounts, including that by Prescott and Pendleton (1952) explicitly identify laterite formations with conditions of planation. This is due in part to the apparent flatness of many residual, exposed laterite sheets, and to the facility with which it is possible to reassemble such fragments into apparently extensive surfaces of low relief. Such ideas are also in accord with most reasoning concerning the development of laterite deposits, for although this is a matter of endless controversy, there is nonetheless some general agreement that laterite formation is to be associated with small oscillatory (vertical) movements of the groundwater table. It is reasoned that such conditions will obtain only beneath stable landsurfaces lacking in steep slopes.

It is clear, however, that several distinct notions are involved in this basic assumption, and some of these have been questioned. In the first place the circumstantial evidence comes principally from an assumed reconstruction of continuously lateritised terrains on the basis of surviving residual hills, usually of the mesa type. However, it is readily seen that, even beneath a surface of planation, watertable conditions will vary as between interfluves and valley floors. If therefore it may be assumed that laterite formation depends on critical relationships of this kind, then it is likely that lateritic horizons have seldom been continuous over large areas. Prescott and Pendleton (1952) recognised the possibility that the surviving residuals represent laterite formations which were formerly located along the lower slopes of gentle valleys; intervening divides which lacked the protective horizon of laterite having since been eroded to form topographic lows in a process of relief inversion. Thus, even in the absence of further theoretical problems, it is not justifiable to associate existing laterites directly with the form of the surface beneath which they were formed.

It is more important to dissociate terrains of low relief and gentle slopes from base-levelled plains. There is no reason to suppose that extensive laterite sheets have developed only on land surfaces at or near base-level. Even if we accept a cyclical framework of thought, it is clear that former planation surfaces will survive for long periods even at high altitudes, until attacked by the encroaching headwaters of incised streams. It seems likely that laterites developed simultaneously across landsurfaces at various altitudinal levels, but perhaps only where conditions of relief were preserved from some former lowland situation. Similar laterite formations found above and below non-faulted escarpments in Nigeria support such an hypothesis. Maud has noted in South Africa (1968) that similar deposits are found on both mid-Tertiary and later Cenozoic surfaces, and he suggests that the laterites may be younger than either of the surfaces on which they are found.

The formation of deep weathering profiles and of lateritic horizons takes a long time, and it is generally agreed that whatever the detailed requirements of climate and terrain necessary for the development of the laterite, any serious disturbance of landsurface stability will halt the process and probably lead to dissection of the laterite or to its burial. A stable landsurface is one over which agents of denudation operate with minimum intensity and is one within which the rate of weathering equals or exceeds the pace of surface erosion. Under a forest cover this may not imply a complete lack of relief. Sivarojasingham *et al.* comment (1962, p.17) that 'high rainfall and high iron content of parent rock may favour the formation of laterite even where the surface is not level but is stable owing to the nature of the weathered material and forest cover'. Several variable factors are emphasised in this passage, and these make generalisations about terrain conditions for lateritisation very difficult. Mulcahy (1964) and Playford (1954) both postulate laterite formation on slopes of up to 10° and with a local relief of 60 m in south-western Australia. Similar conditions appear to obtain in Uganda (Trendall 1962; De Swardt 1964; De Swardt and Trendall 1969). But in this case Trendall and De Swardt subscribe to the view that the laterite has continued to form and thicken during a period of lowering of the land surface throughout which the surface soil was maintained.

This hypothesis, which is considered further below (p. 67), dissociates laterite formation from planation, but conceives the persistence of a steady state of low relief during a prolonged period of uplift and denudation. The flatness of the interfluves in dissected laterite terrains is probably induced by the lateritic horizon itself as Trendall (1962) comments, and is thus not a part of the original landsurface. If all this appears inconclusive, it should not be thought that the latitude within the argument is great. It seems unlikely that primary laterite deposits will develop extensively unless there is a degree of stability in the denudation system of an area. This will hardly obtain if relief development is great or slopes steep. On the other hand there is considerable field evidence to suggest that, once a thick duricrust has formed, conditions for the re-formation of lateritic horizons and surface layers may persist even during phases of active relief development, and on steep slopes of up to 20° or more.

The origin of iron in laterites

A question was posed earlier in the discussion concerning the great thickness of many laterite sheets. Thus, although few laterites exceed 2 m in thickness over wide areas, depths of 12 m are not uncommon locally and much greater thicknesses have been claimed. If aluminous laterites are excluded from consideration it is common to find that lateritic horizons contain from 30 per cent to as much as 80 per cent iron oxides (Fe)O)/FeO). On the other hand unaltered granite will contain in the region of 2 per cent of iron oxide (Trendall 1962), whilst a more basic rock such as dolerite may contain as much as 11 per cent of combined iron (Sivarojasingham *et al.* 1962).

Ultimately all the iron in the laterite must come from such sources, but its concentration into a distinct horizon poses many problems, and one of these arises because a laterite deposit will require the alteration of many times its own thickness of original rock to supply the necessary iron. According to Trendall (1962), the factor for granite in Buganda is about 14 times, but since some of the iron would be lost, this rises to a factor of 20 times. Thus to supply enough iron for 9 m of laterite some 180 m of rock would have to be weathered, if all the iron were to come from *in situ* alteration. We probably cannot postulate the widespread occurrence of 180 m of weathering, and if we could, it would not be possible to argue for the concentration of all the iron from such a deep profile into a narrow horizon near the surface. But some long range upward movement of iron may in fact occur, according to an hypothesis advanced by Lelong (1966), whereby ionic diffusion may remove products of rock alteration from deep within the weathering profile (see chapter 1, p. 31). According to Lelong, iron precipitation within 2 m of the surface could arise in this way. The hypothesis remains unproven, but is far reaching in its implications. This situation only arises, however, if it is thought necessary to derive the major part of the iron in the laterite from the site of formation. Trendall (1962) considered this to be the case, but few other writers have reasoned in this manner. Most have suggested that iron is derived from other sources, beyond the site of formation, and that iron-bearing solutions move laterally across the landscape. As the iron becomes immobilised in favourable sites, so considerable thicknesses of laterite accumulate.

This situation was recognised by Sivarojasingham *et al.* (1962), who defined the problem in terms of the differences in iron content observed between the weathered rock (saprolite), soft laterite and hard laterite. The following figures come from observations of granite weathering at Samaru in Nigeria. Only the principal constituents are shown here:

TABLE 8

VARIATION IN MINERALOGY WITHIN A LATERITE PROFILE AT SAMARU, NIGERIA

Mineral	*Granite saprolite Depth*	*Soft laterite Depth*	*Hard laterite Depth*
	60–120 in.	32–64 in.	0–32 in.
SiO_2	67·3	55·2	35·6
Al_2O_3	18·4	21·4	18·8
Fe_2O_3	6·3	13·7	33·9
H_2O	7·1	8·9	10·1
Kaolin	40	30	15

(After Sivarojasingham *et al.* 1962)

If it could be assumed that the actual amount of iron remained constant, then, in the above example, the net loss of silica would be 90 per cent, and of alumina 81 per cent. These figures are much too high to be explained by the weathering of kaolin in the saprolite, and most of the rest is quartz. It is therefore necessary to postulate the addition of iron from other sites. There are some laterites where the Fe:Al ratio is identical to the original rock; in these cases it is unlikely that iron has been added from outside.

Iron becomes mobilised in weathering profiles as a result of reducing conditions, and the presence of organic complexing (chelating) compounds. These conditions arise under forest (and perhaps woodland) vegetation, and particularly under wet season conditions of high soil humidity and leaching. Conditions of high acidity will also mobilise the iron, but these seem confined to swamp environments Bleackley (1964). On the other hand repeated wetting and drying, particularly under conditions of low organic content of topsoil, will lead to the precipitation of the iron. Mobilisation of the iron appears to be most marked in the pallid zone, where long periods of saturation occur and the drying out of the regolith is rare. Much of the iron released in this zone probably moves either downward or laterally. The lateral movement of the iron results in deposition near plateau edges, where very thick laterites are commonly found (see Maignien 1966; Brosch 1970).

Enrichment of the saprolite implied by the figures in table 8 can thus arise in the following ways:

(1) Addition of iron from above the developing laterite horizon. Percolating water from the thin topsoil may carry small amounts of iron and aluminium into the laterite but they must be regarded as insignificant, unless the topsoil has been enriched by iron moving laterally within the profile (see below).

(2) Capillary rise of soil water carrying iron may occur over a narrow zone, but it is a slow process, and likely only to occur under dry season conditions. Iron added from this source is likely to be in very small amounts.

(3) Migration of iron from greater depths could possibly take place by ionic diffusion, during the drying out of the regolith in the dry season (Lelong, 1966). Such an hypothesis would account for the removal of iron from a thick pallid zone and its concentration within 2 m of the surface, where desiccation of the regolith occurs.

(4) Many writers regard the site of iron precipitation as the upper level of the water-table. The water-table relationships in permeable regoliths are such that above the zone of permanent saturation within which oxygen is largely excluded, there is a zone of intermittent saturation (vadose zone) above which, in the surface soil, conditions are seldom if ever saturated. If this vadose zone is narrow, then iron can become concentrated within it to form a lateritic formation: small quantities being deposited at the upper limit of saturation as it varies in height within the profile.

(5) Iron-bearing solutions may pass laterally through the soil to enrich the profiles beneath the lower slopes of valleys. Iron is thus leached from regoliths beneath interfluves and deposited on gentle slopes below higher land. This process becomes accentuated if iron may also be derived from higher level laterites undergoing fragmentation and dissolution. This last source of iron is considered to be of major importance by D'Hoore (1954) and Maignien (1958, 1966; and figures 13 and 14).

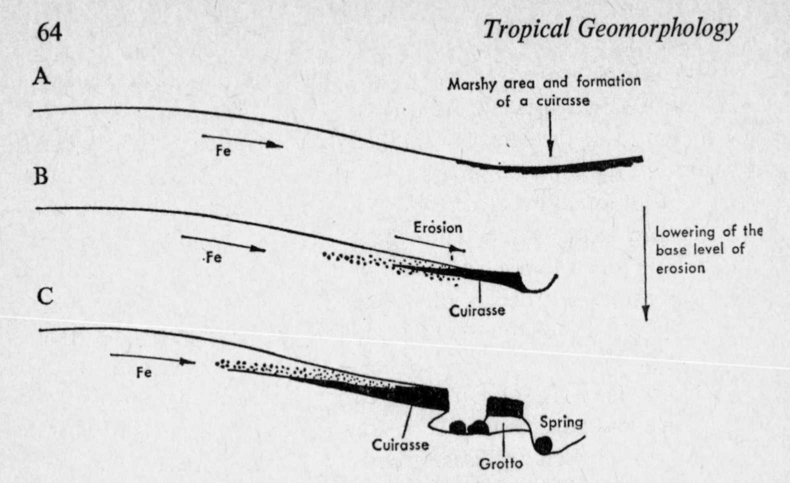

Figure 13 Migration of sesquioxides and formation of a laterite cuirasse (from Maignien 1966).

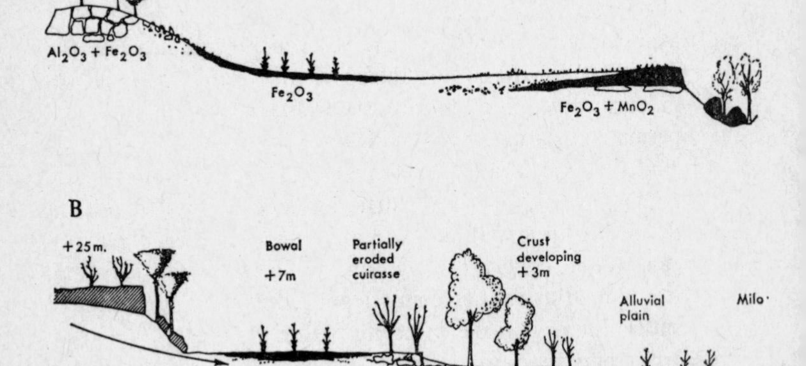

Figure 14 Association of laterities with relief features (from Maignien 1966).

A. Catena showing denuded ironpan; colluvial ironpan; and ironpan formation on terrace. Near Labe, Guinea.
B. Catena showing three levels of crust formation on terraces of the Milo River, Guinea.

(6) Enrichment of iron by removal of other constituents has been shown to be insufficient in the case of granites, but laterites may result from such a process in the case of basic rocks. This process of relative enrichment was thought by Harrison (1934) to be responsible for the development of true Primary Laterites on elevated sites in British Guiana (Guyana). Some of this reasoning is followed by Bleackley (1964), but it appears that in most cases such deposits are aluminous laterites, often meriting description as bauxite.

It is difficult to evaluate the importance of these different mechanisms. If relative enrichment is regarded as important mainly in the case of aluminous laterites developed from basic rocks, then mechanisms 4 and 5 appear likely to be quantitatively the most important in the development of laterites. However, the process of ionic diffusion described by Lelong (1966) is attractive because it also offers a mechanism for the development of closed depressions in the basal surface of weathering from which solutes including iron appear unable to escape by alternative means. The emphasis given to one or the other of these processes affects many arguments of a geomorphological nature and they will thus be discussed further.

Enrichment by lateral movement of iron

Lateral movement of soil water is favoured by relief development, but this also is likely to extend the range of water-table fluctuation in such a manner that the concentration of iron from this cause will be inhibited. However the watertable fluctuations will remain small on the lower slopes of valleys, and where these are wide and gentle extensive laterite deposits may form, as iron is added from adjacent, high ground.

Mohr and van Baren (1954) comment that at the base of the vadose zone there is commonly a horizon of kaolin concentration which may seal the profile, promoting the lateral movement of soil water above this layer. This process may be assisted by the existence of a highly permeable stone-line (Berry and Ruxton 1959). These are two general considerations which suggest that lateral enrichment may be a normal process in laterite formation. But two further situations add force to the argument. First, additional iron may be derived from the retreat of slopes around older laterite capped profiles. Field observation emphasises that these older caps undergo fragmentation, giving rise to lateritic gravel which may become recemented on lower slopes. However, in addition to the fragmentation affecting the margins of laterite plateaux in savanna areas, it is generally agreed that laterite deposits may be redissolved. Both D'Hoore (1954) and Nye (1955) refer to this process, and it is most likely to occur beneath a forest cover, probably from unexposed, soft laterite formations. It is therefore a false argument to suggest that only detrital laterites can form from the addition of iron derived from older and higher deposits.

A broader view of the geomorphological development of tropical terrains may add a further dimension to this process. Ruxton and Berry (1961a) have commented that the back-wearing of major hillslopes may contribute important quantities of iron to the profiles in the surrounding zone: 'on the back-wearing theory only a minute proportion of the total iron in a very large quantity of rock waste removed from the hillside is required to provide the enrichment' (p. 29). The iron might be added in solution or, more particularly, as a coating on kaolinite clay particles added to the soil by mechanical eluviation.

It thus appears that many circumstances favour the lateral addition of iron into developing laterite profiles, and it is probable that these account for the occurrence of particularly thick formations. It should be added here that Maignien (1958, 1966) particularly, has commented on the thickening of laterites towards exposed faces, and conversely on the fact that laterites commonly thin rapidly from such faces, and may disappear in the interior of large plateaux. This 'corniche' laterite may be due to the lateral movement of water along the impermeable mottled clay layer and the rapid precipitation of the contained iron close to the exposed face, where drying out occurs and organic matter is at a minimum (figures 13, 14). The fact that lateritic horizons themselves produce conditions for the immobilisation of iron is a further factor leading to the progressive thickening of existing sheets in favourable locations. It is also likely that these will be preferentially preserved in the landscape following dissection, owing to their superior resistance to erosion.

Enrichment by water-table fluctuations

The concept of a rising watertable bringing dissolved iron towards the surface from some lower site of rock weathering, encounters major objections which have been clearly stated by Maignien (1966). Although the watertable rises and falls within weathering profiles it is not clear how upward movement of water could occur, except perhaps by hydrostatic pressure from above. After prolonged rainfall, the upper vadose zone will contain 'new' water which has passed downwards to rest on the 'older' groundwater below. Maignien thus concludes that 'the upper layer of rising groundwater table should carry no more material for enrichment than is dissolved during its descent through the overlying soil and laterite zone'. The corollary of his reasoning is that 'we can admit the operation of segregation mechanisms due to fluctuating water tables without appreciable overall enrichment' (1966, p. 101). This argument however, again ignores the work of Lelong (1966), who suggested that upward transfer of both water vapour and dissolved irons may occur during the dry season, thus effectively transfering 'old' groundwater towards the surface, where it may evaporate or become subject to lateral movement.

Trendall (1962), however, takes a quite different line of reasoning. Because the rise in the water-table results from the addition of unsaturated water from above, the most active weathering and the release of iron take place within the vadose zone during the wet season, with the concentration of the iron at the upper surface to form an embryonic laterite horizon of minimal thickness. With the onset of dry season conditions a gradual fall in the water-table leads to the removal of salts still in solution to lower levels. Consequently there is a settling of the upper part of the profile containing the laterite. If, at the same time, a gradual lowering of local base levels takes place through stream activity, a similar lowering of local water-tables will occur. If the subsidence of the profile due to removal of solutes and the lowering of stream channels continue at similar rates then, as the laterite horizon subsides, so further iron will be added from below. Thus, iron can be derived each wet season from progressively lower zones of the parent rock. In this way, according to Trendall (1962), thick laterites could be formed from considerable thicknesses of rock, without the addition of laterally moving, iron-bearing solutions. It is inherent in this hypothesis that the formation of the laterite deposits has occurred during a prolonged period of slow down-wearing, during which both stream channels and interfluves have been lowered at a similar rate, maintaining steady state landforms. In fact Trendall (1962) calculated that to produce a laterite crust of 9–10 m in thickness, the landsurface would have to be lowered by 180 m without marked morphological change. Geological evidence is cited to suggest that this process may have taken 35 million years.

Such an hypothesis encounters many objections. In the first place the assumption that important volume changes occur during weathering has been criticised by Ollier (1967, 1969). But even if the basic process is allowed, its persistence over such a time scale is unlikely. Intermittent crustal rise (Wayland 1934; Schumm 1963) is likely to have profoundly affected rates of channel erosion which generally exceed rates of hillslope erosion during uplift (Schumm 1963). Moreover, major climatic oscillations have occurred during the Tertiary period as well as in the Quaternary, and these have had the effect over many areas of removing topsoil to expose hardened laterite sheets which become subject to marginal attrition through lateral erosion. The probability that iron has been derived from other sources than the underlying rock further weakens the argument.

However, although it has been found necessary to criticise the general hypothesis put forward by Trendall, some of the issues which his paper poses remain important, and the possibility that laterite deposits may subside or may in other ways extend downwards into the weathering profile is of some significance to the present argument. For instance Nye (1955) suggested that iron dissolves from the top of the laterite and enriches the underlying layers as normal erosion lowers the zone of saturation. If this process operates widely, and it would presumably be confined to forested areas, it may add credibility to certain aspects of Trendall's main hypothesis. It also accords with the general hypothesis for soil formation advanced by Nikiforoff (1949, 1959).

In summary, the iron content of laterites appears to have several possible sources, the most important of which may be listed:

(1) From *in situ* weathering within the vadose zone, possibly with a lowering of this zone and the overlying lateritic horizon with time.

(2) From neighbouring rocks on higher sites and brought into the profile by laterally moving groundwater.

(3) From older and higher laterites by fragmentation and solution, and brought into the profile by laterally moving groundwater and by surface wash of detrital fragments.

(4) From the permanently saturated pallid zone by upward ionic diffusion, arising from a gradient of concentration in the groundwater.

Whatever the source of the iron, it appears to be deposited most commonly at the upper surface of the water-table where oxidation of reduced ferrous iron in solution readily occurs. The situation is rarely as simple as this however; the iron is immobilised not only by simple oxidation but also due to the destruction of organic complexing agents, and by adsorption on to existing ferric compounds and kaolin. The iron may be immobilised as a result of laterally moving water approaching the surface, leading to evaporation and oxidation of the iron; as a result of discontinuities in permeability, as well as a result of normal watertable relationships.

The induration of laterite formations

Although the hardening of laterite is associated with exposure, due to the stripping of topsoil, or to the outcrop of the lateritic horizon on valley sides, the process involves crystallisation of the iron oxides as well as simple dehydration. The iron content appears to be the key to this process and Maignien (1966) shows that the harder laterites are found to have a higher iron content. He concludes (p. 41) that the 'iron-bearing parts of the hardened laterite have a greater degree of crystallinity, or a greater continuity of the crystalline phase, than the soft materials with which they are associated'. There also appears to be some further leaching of kaolinite which adsorbs iron and prevents the formation of crystalline goethite. It follows that both wetting and drying are necessary to bring about hardening, and Maignien comments that hardening is not obvious either where rainfall is low or where it is almost continuous. Vegetation removal or degradation due to climatic change would expose the surface soil to extremes of temperature and to rapid cycles of wetting and drying after heavy rainfall. These conditions would also promote rapid runoff and the stripping of overlying soil horizons thus leading both to the exposure and induration of the deposit.

Incision of streams into a lateritised weathering profile may give rise to local induration around the valley, causing 'gallery' laterite benches to form, even under a mature rainforest vegetation.

The development of secondary laterite formations

So long as an assumption is made that iron in primary laterites is derived mainly from below the site of formation, it is also easy to distinguish between this type of formation and one which results from the break-up of a primary laterite sheet and the recementation of the fragments at a lower level in the landscape. However, it has been shown that lateral movement of iron in solution probably plays a major part in the formation of most laterite sheets, and many which may be developing at the present time on valley flanks below older fossil crusts, may not in fact owe their formation so much to the fragmentation of the higher sheets as to the remobilisation of iron from the older laterite, and also from intervening zones of the weathering profile. The lower laterite may therefore result in part from the same processes as gave rise to the older crusts.

However, in specific circumstances a detrital origin for part of the material in developing laterites may be unambiguous, and in such cases the terms secondary or detrital laterite can be used. Such laterites appear as type C in table 6 (p. 52), where indurated elements cement pre-existing materials. Their morphology is often described as concretionary, and they may take on conglomeratic, brecciated, or nodular forms containing large fragments of old crusts together with quartz, sand and gravel. The cementing process, however, is one of lateritisation and the fragmentary material is often incidental to the formation of such deposits. Maignien (1966) claims that such detrital deposits are less common than the literature suggests and that dissolution of sesquioxides in older formations becomes a more important process than mechanical breakdown, wherever the annual precipitation exceeds 500 millimetres. It is possible to doubt this assertion, but at the same time agree that long distance lateral transport of the fragmentary material is uncommon. Both Maignien and Moss (1965) point to the fact that the undermining of older duricrusts results in settling of large blocks and smaller fragments which may later become recemented more or less *in situ,* or may dissolve to provide iron in solution which migrates to sites more favourable for precipitation (figure 14).

In fact a close look at the distinction between primary and secondary laterites makes it quite clear that its validity is doubtful. Sivarojasingham *et al.* (1962, p. 54) emphasise 'that mixing and translocation of superficial materials appears to be the rule rather than the exception on old land surfaces'. They point out that our understanding of such surfaces as being due to extensive planation, implies the former existence of still higher surfaces and that with the destruction of these, some inheritance of material is bound to take place. They conclude that 'if viewed in this perspective, it is reasonable to expect that many of the high level forms were subjected to the same kinds of mass movement, lateral drainage, and enrichment observed on low-lying erosion surfaces today'.

Thus the distinction formerly thought to be genetic may reflect only the chronology of laterite development. The widespread recognition of an older and higher laterite forming tablelands, and a younger and lower laterite occurring as valley-side benches in tropical Africa (Du Preez 1949; De Swardt 1964), can perhaps be seen in this context. However, it may be anticipated that the younger laterites will contain a higher proportion of detrital fragments than older deposits.

The convenient classification of laterites into two main classes is in itself a simplification, and several different levels of accumulation may be evident in a single landscape (Sombroek 1971). Maignien (1966, p. 68) points out that 'the intensity of ferruginous incrustation reflects the capacity of a catchment basin to supply a certain amount of iron which accumulates in subhorizontal sectors which function as reception zones. Therefore any subhorizontal form is capable of becoming incrusted.' Whether detrital material is present in large quantities or not will depend largely upon the geomorphic situation and sequence of development.

Recent studies in south-western Nigeria by Fölster (1969) appear to confirm this concept. He found, in accordance with earlier writers, that laterites can be divided between summit patches, and lower occurrences on pediment slopes. These latter are 'restricted to the proximity of drainage depressions. Their frequency as well as their extension increases with decreasing size of the stream and reaches its peak in the small drainage dells on the pediment slopes' (Fölster 1969, p. 24). Thus, recent and current laterite development appears to be closely related to convergent seepage flow, and to be most effective where evacuation of water via drainage channels is comparatively feeble and restricted to short periods at the height of the wet season (figure 13).

Distribution and origins or related deposits

Bauxite. Aluminous crusts or bauxites appear to form under conditions of free leaching where rainfall exceeds 1200–1500 millimetres. The concentration of alumina results from the leaching of most of the other constituents including silica and iron. But while this suggests a different set of environmental conditions to those operative for laterite formation, the transition from ferruginous to aluminous duricrusts often occurs locally and within a single deposit. However, the best developed bauxites are often associated with one or more of the following conditions:

(1) A humid tropical climate with a forest vegetation which promotes organic complexing of iron and its removal from the profile.
(2) Elevated sites where free leaching by percolating groundwater is promoted.
(3) Rocks rich in aluminium (and poor in iron), such as basic volcanics and phyllites.

In detail, however, the situation is complex. Two principal theories of bauxitisation are commonly offered. In the first, primary lateritisation is held to give rise to a deposit rich in gibbsite which develops directly from the parent rock. This was put forward by Harrison (1934) and has been supported by Bleackley (1964), who points out that mineral structures are commonly preserved in gibbsitic laterite crusts in Guyana, although they are underlain by deep and structureless clays. He argues that this situation requires gibbsite formation during an initial phase of rock weathering leading to crust formation, and that re-silication has resulted in clay synthesis and the loss of mineral structures during subsequent weathering below the water-table. Gradual lowering of the water-table during stream incision could have led to increasing thickness of the primary, bauxitic laterite. However, the bauxites of commercial grade have clearly resulted either from a further loss of iron from the primary laterite, or by de-silication of the kaolinitic clays. Bleackley (1964) describes an arcuate belt of tabular, bauxite ore bodies that appear to mark a former strand line extending from Guyana southwards into Brazil. He argues that the primary, ferruginous bauxites, having lost much of their iron through chelation and related mechanisms, became redistributed by erosion and accumulated in lagoonal depressions, where they were first covered by later sediments and later uplifted to their present positions.

On the other hand bauxite formation by de-silication of kaolinite is favoured by many writers, but the mechanism is not clear. Nevertheless there is much evidence to show that bauxite formation occurs generally within the freely leached zone above the water-table, and some evidence has been offered to suggest that it may grade down into more ferruginous material at depth. However, lateral variations are no less apparent (Grubb 1968). Both rock composition and rock texture may also be important, and Grubb (1968) suggests that silica loss and bauxite formation are most evident in fine grained rocks.

In geomorphological studies the possibility that bauxite deposits may form and continue to thicken during periods of relief development is important and may offer a contrast to laterite formation.

Silcrete. In contrast, silica rich duricrusts are associated with semi-arid conditions. But agreement is lacking concerning the formational histories of many such deposits. It has been suggested that siliceous duricrusts develop where precipitation declines below 250 mm, and there is little doubt that in Australia, for example, the gross distribution of silcretes is to the arid side of the laterite formations (Langford-Smith and Dury 1965; Aitcheson and Grant 1970). But silica mobility is known to be increased by alkali conditions which are not only associated with low rainfall. Depressions in the landscape, where drainage is poor, commonly exhibit high pH values for groundwater, and silica may be mobilised.

According to Frankel and Kent (1937) silcretes may develop as a result of the upward migration of colloidal silica, released by leaching at depth in the weathering profile where, (as noted in chapter 1) alkali conditions may obtain. Precipitation of the silica, according to these authors, may result from coagulation due to the meeting of saline solutions percolating from the surface downwards. The formation of the silcrete horizon would therefore depend upon delicate adjustments in the water-table of the profile in a manner analagous to that suggested for laterite. Other writers such as Mountain (1951) consider silica to migrate downwards through the profile as a result of the further weathering of clays.

The association of silcrete formation with the lower, pallid, zone of a laterite profile could be accommodated by these ideas, although it has been seriously criticised by Langford-Smith and Dury (1965).

If silcrete is regarded as a primary formation, and part of a complete and largely *in situ* weathering profile, then the conditions favouring its formation may be:

(1) A semi-arid climate with precipitation in the region of 250 mm.
(2) Depressions in the landsurface, where poor drainage promotes high pH in groundwater.
(3) Areas of strong sheet wash, where silica is transported as a gel, but is also deposited when the water evaporates.

However, many silcretes have been interpreted as secondary modifications of formerly complete laterite profiles. Broadly, silcretes (excluding sub-basaltic quartzites) have been interpreted in one of three ways:

(a) As the upper horizon of a complete profile, in which case the criteria discussed above apply.
(b) As a lower horizon of a truncated laterite profile, in which case a particular geomorphic relationship, as well as a climatic one is implied.
(c) As a result of the silicification of laterite, perhaps due to climatic change.

These may not be mutually exclusive possibilities. However, in both the case of bauxite which is associated with humid climatic conditions, and silcrete which is associated with semi-arid conditions, it is scarcely possible to conclude that the climatic control is always clear and unambiguous. However, the low pH values maintained by the leaching active in humid climates would appear to preclude the concentration of silica except in unusual sites or over alkali-rich rocks. Iron, on the other hand, is removed both in alkaline and in slightly acid conditions. Some complex profiles from northern Australia have been described by Hays (1967).

Some writers have emphasised that silcretes are mostly associated with waterborn deposits within which the silica was present as a gel.

LATERITE AND LANDFORM DEVELOPMENT

The conditions of terrain necessary for the development of laterites have as their corollary the effects which laterite formations have upon the landform. As has been shown, this is a complex topic, for while laterite formation will only occur if certain limiting conditions of terrain are available, the existence of laterite formations undoubtedly influences the development of landforms in important ways. The discussion can be considered under three main headings: (1) landforms associated with laterite deposits, (2) slope development within the laterite profile, and (3) landform development and evolution in laterised terrains.

Landforms associated with laterite deposits include the following:

(1) Tablelands developed on flat or subhorizontal laterite sheets which may vary in extent from small mesas to extensive plateaux (see plates 1, 2).

(2) Cliffs or 'breakaways' marking the edges of the tablelands (see plate 3). These may degenerate into linear slopes of laterite rubble.

(3) Bench or terrace-like features on intermediate slopes of valleys, usually marked by a break of slope on the downslope side.

(4) Pavements of recemented laterite fragments forming on lower slopes and valley floors.

(5) Circular or eliptical hollows varying in dimension from a few square metres to many hundreds of square metres. These have been given local names such as 'Bowal' (Australia) or 'Baixa' (Guyana).

In addition to these one might list slopes dominated by laterite fragments. These may take three principal forms: (1) talus or boulder controlled slopes; (2) pediment slopes carrying a superficial cover of pisolitic gravel; in arid areas these may have negligible slope and become regs; (3) colluvial slopes, probably restricted to forest areas as described by Moss (1965); see plate 8.

Certain of these forms occur most commonly in combination, as illustrated by the accompanying diagrams in figure 15. The depressions, or hollows, are found on the surface of the laterite sheet which is continuous beneath them, while the margins of the duricrust are commonly marked by cliffed breakaways below which boulder controlled slopes give way to pediment features, cut across the lower zones of the laterite profile. It is usually on the lower portion of this pediment that lateritic material reforms, often cementing the detritus from the higher sheet. According to the valley form, these secondary or lower laterites produce pavements or bench-like features.

Slope development on such a landform appears to follow a simple pattern, but some variation results from the tilt on the laterite sheets on the one hand and from differences of climate and vegetation on

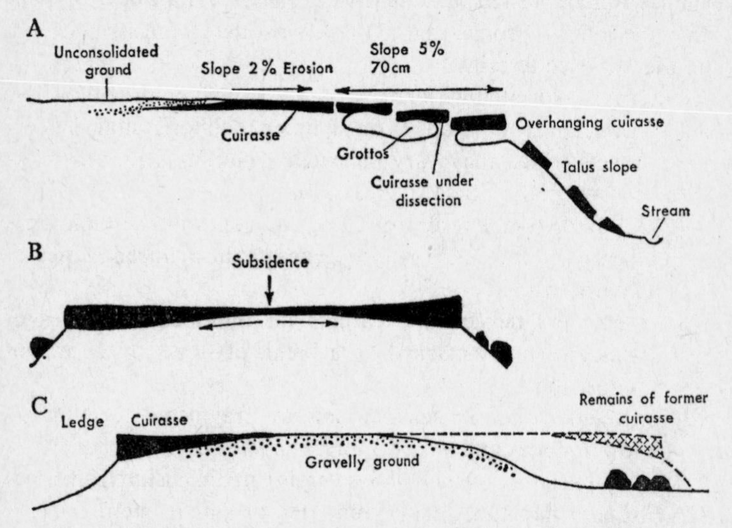

Figure 15 Some aspects of the morphology of duricrusted hillslopes (from Maignien 1966).

the other. Basically, the indurated laterite sheet undergoes a form of parallel retreat, leading to the extension of a pediment slope across lower zones of the weathering profiles. Pallister (1956) has clearly shown that slope retreat takes place and that with the final disappearance of the laterite sheet more gentle convex slopes form and the residual hills may undergo a period of slope decline (figure 16).

This type of slope evolution depends in part at least upon the resistant nature of the indurated laterite on the one hand and upon the impermeable character of the underlying mottled clay zone on the other. Although observations concerning the porosity of indurated laterite vary considerably, there is little doubt that water can infiltrate these formations quite rapidly, either via interconnecting channels or through joints which develop in the cuirass. Spring

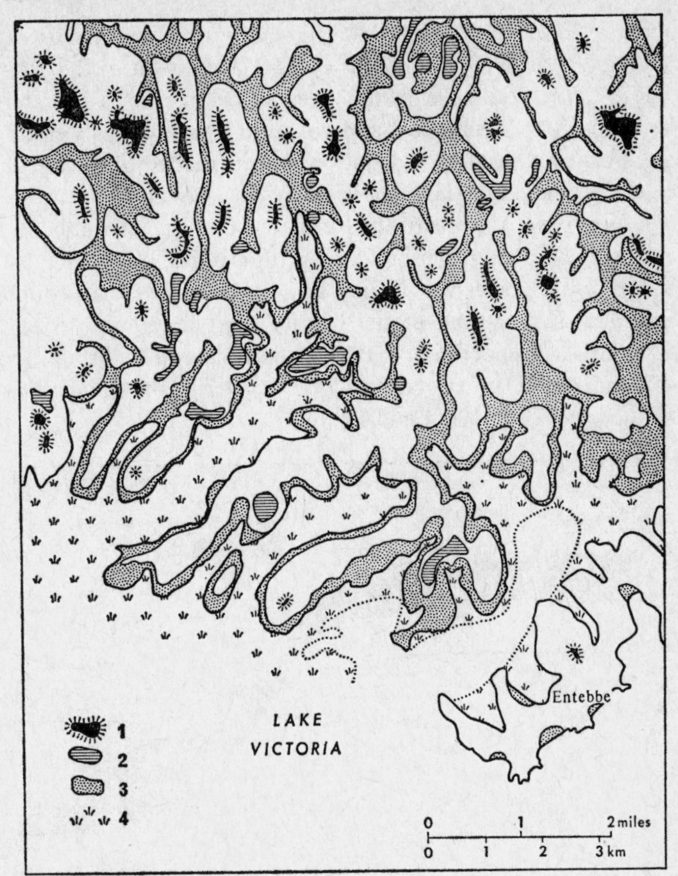

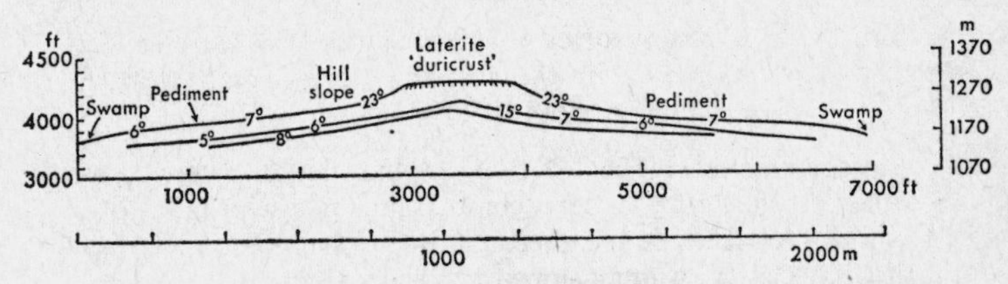

Figure 16 Landforms associated with laterite duricrust in part of Buganda (after Pallister 1956)

1. Flat hill-summit with laterite duricrust, surrounded by slopes up to 24° (mid-Tertiary surface?); 2. Flats (up to 5° inclination) associated with later (end Tertiary) surface; 3. Alluvial silts and lacustrine beds; 4. Papyrus swamp.

sapping at the base of the laterite has been noted by many writers including Vann (1963) and Moss (1965). The action of the springs or more general seepage from beneath the laterite is to remove material by both mechanical and chemical eluviation as well as by straightforward flushing. Cavitation within the mottled clay zone may occur and quite large hollows result. Collapse of the crust above these caverns leads to the formation of large laterite boulders which may undergo further breakdown into rubble, and subsequently into a pisolitic gravel which may be washed across the pediment. A more gradual settling of the laterite crust may occur, where a slow removal of underlying material takes place. This brings about cambering of the outer margins of the laterite tablelands clearly illustrated by Moss (1965) and figure 17.

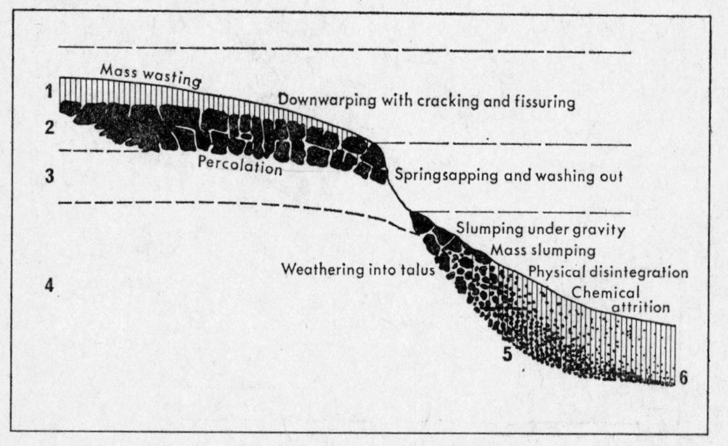

Figure 17 Processes involved in the retreat of laterite (plinthite) breakaways, in the forest zone of south western Nigeria (after Moss 1965).

1. Surface mantle; 2. Hard duricrust (plinthite); 3. Altered and weathered sandstone; 4. Ferruginous sandstone with mudstone bands and lenticles; 5. Weathering talus; 6. Non-mottled ferrallitic sandy clay on waning slope or pediment.

Many of the laterite caps have slopes of 1–2° which may be due to the attitude of formation or to differential settling similar to that already described. Once this tilt has been established, however, it leads to a basic asymmetry of form on the residual hill. Groundwater is channelled downslope and the cavitation and collapse of the crust becomes most evident on this side, whilst on the upslope side a more gradual crumbling of the crust leads to a boulder controlled slope, commonly of around 15 degrees.

A difference in process and form exists between the savanna areas and the forest, and in some cases the nature of the underlying rock also influences the development of slope forms. In the savannas, and in semi-arid areas, wash slopes or pediments form across the pallid or transitional zones of the weathering profile, and the laterite fragments are carried across this slope as a pisolitic gravel. In more humid regions, colluvial processes become more important. In south-western Nigeria, Moss (1965) describes the formation of a non-mottled, ferrallitic sandy clay below the retreating laterite face, as a result of marked vertical slumping but little horizontal movement of the lateritic waste (figure 17). If this detritus is not removed by creep and wash processes below the breakaway, it may build up to obscure the laterite cliff and lead to a more subdued form. Within the forest zone this development may be quite common as a result of the more humid climate since the last sub-humid phase of the Pleistocene.

In areas of high but seasonal rainfall, and over basic rocks as in parts of central Sierra Leone, more complex slope forms exist. In the Sula Mountains a lateritised surface with a high internal relief is found. Marginal slopes are characterised by slump scars and recemented earthflows of lateritic rubble (figure 18). In this area the basic schists have yielded a clay-rich regolith capped by massive laterite deposits. Intense rainfall in the wet season (750 mm in August alone) saturates and overloads the regolith which flows or slumps on the steeper slopes, carrying the overlying laterite with it. The iron content of the groundwater here appears to be sufficient to lead to recementation of the rubble even on slopes of up to 20°, presumably in response to the equally severe dry season of some four months' duration.

Enclosed hollows or 'baixas' have been described from many lateritic surfaces. In Buganda (Pallister 1956) and Guyana (Sinha 1966) these hollows are commonly 3–30 m across and some 3–4 m deep. They have been ascribed to settling or foundering of the crust due to solutional removal and eluviation of fine material from below. A much larger feature containing Lake Sonfon occurs in Sierra

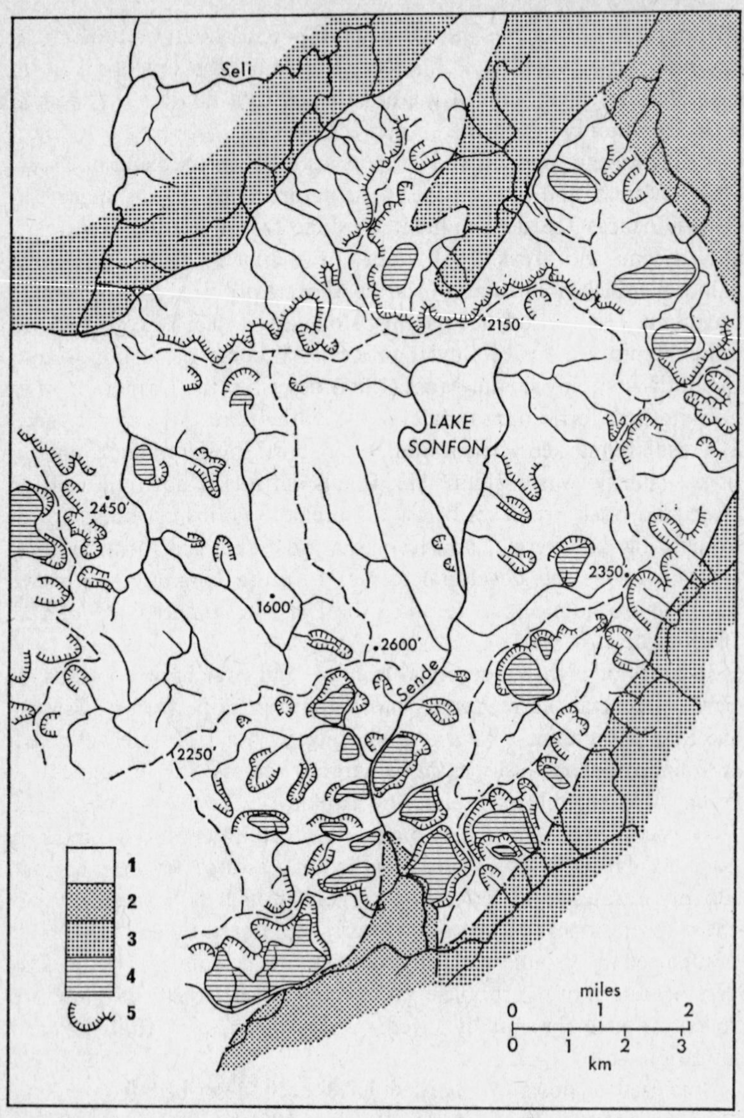

Figure 18 Lateritised landsurfaces of the Sula Mountains, Sierra Leone.

1. Duricrusted surface over basic metamorphic rocks (mainly amphibolite, hornblende, talc and chlorite schists); 2. Granite; 3. Gneiss and migmatite; 4. Earthflows; 5. Slump scars.

Leone. This depression is several square kilometres in extent and some 60 m deep. As described, this is morphogenetically a very active zone, but whether it could have arisen from the operation of similar processes has been doubted. Wilson and Marmo (1958) suggested that it may have resulted from drainage reversal following crustal warping, but little evidence was adduced to support this notion. Both the development of the internal depressions and the cambering and collapse features of laterite terrains indicate that, particularly under humid conditions, the laterite crusts may undergo differential movement and settling. It is possible that the marked slope on some sheets is a result of similar processes, accentuating some small original inclination of the laterite formation.

The settling of laterite sheets forms a major part of Trendall's hypothesis (1962) of landform development in laterite terrain. Trendall requires a general lowering of the laterite as material from the underlying pallid zone is removed in solution. This has been challenged by the implication that breakaway retreat will operate more rapidly than any gradual lowering of the profile, so that slope replacement and the development of secondary laterite formations will be more evident than a persistence of steady state forms. This argument receives support from observations of terrains where settling of the laterite sheet can be inferred from field evidence, for this process implies active morphogenesis and the production of new forms or at least major modification of older ones. Such observations suggest that mass movement and slope failure are much more important processes than solution in the development of laterite terrains. Nevertheless, the effects of solution may also be important.

It is of major significance that laterites tend to bring about relief inversion or at least the survival of high level surfaces. On a local scale laterite formations which may have originated beneath lower slopes of valleys, are now commonly found as summit cappings forming ridges, plateaux or small mesas. This implies the differential lowering of those parts of an original landscape which were not protected by the laterite: a sequence which probably began with the incision of streams, but may also have involved the lowering of earlier interfluves not protected by thick duricrusts (figure 43).

On the regional scale, laterite terrains are often observed to stand out as high land although they cap deeply weathered, parent rocks. Adjacent acid rocks giving rise to inselberg landscapes commonly form lower land, although individual hills (usually bornhardts) may rise to the level of the laterite terrains. Many instances of this kind can be found in West Africa. The Sula Mountains form an undulating plateau that rises above adjacent granite and gneiss country which contains some remarkable bornhardts in the Gbenge Hills. In Ghana the aluminous laterites capping the Birrimian ranges and the Voltaian Sandstone escarpment form high ranges overlooking neighbouring lowlands which may be 300 m below. Similarly in western Nigeria the highest areas along the watershed between the Atlantic and Niger drainage systems are found to be amphibolitic rocks carrying a thick duricrust. This circumstantial evidence argues against effective lowering of the entire laterite surface. Whatever morphological changes have taken place and continue within the laterite terrains, it is clear that surfaces unprotected by the duricrust are lowered more rapidly unless they form bare, rock hills. This observation which has been made by several authors (D'Hoore 1964; Thomas, 1965b; Maignien 1966) is of importance to any model of landform evolution in laterite terrains and will be discussed further (chapter 9). The patterns of superficial deposits which result from the destruction of laterite sheets are also considered elsewhere (chapter 5).

3 Deep Weathering Profiles and Patterns

The spatial patterns of weathering can be considered at widely differing scales of enquiry: from the concentric zoning of weathered products around a single corestone (Nossin 1967) to the latitudinal variations in weathering depths (Strakhov 1967; figure 1, p. 4) and weathering products (Pedro 1968; figure 2, p. 5) according to gross variations of regional climate. In this section, however, the weathering patterns will be considered primarily in relation to geological and landform patterns. It is necessary first, to consider the characteristic morphological features of commonly occurring weathering profiles.

The character of weathering profiles

There is no single type of profile illustrating the transition from thoroughly weathered material to fresh and unweathered rock that remains valid for all rock types or all locations. Roe (1951, p. 53) gives three common types of transition found in Malaysian granites. The rock may weather '(i) to form core boulders, (ii) to form clayey soil and soft rotted rock, which is separated by a well defined line of contact from hard fresh granite, (iii) to give a gradual transition from clayey soil to fresh rock'. Although pertaining to granites, these three possibilities apply with variations to most if not all weathering profiles in crystalline rocks. Ruxton and Berry (1957), Berry and Ruxton (1959), Wilhelmy (1958) and Brunsden (1964) have described in detail granite weathering profiles of type (i) above (see plates 4, 5).

The granite weathering profile has become a standard against which to measure other types, but this has in some cases tended to obscure the variations in character between weathering profiles. The standard description given by Ruxton and Berry (1957, 1961a) is illustrated in the accompanying table and diagram (table 9, figure 19). The zones may be briefly described; and sample figures of thickness from Hong Kong are also given:

TABLE 9

CHARACTERISTIC WEATHERING PROFILE OVER WELL JOINTED GRANITE

Soil migratory layer		1–3 m (3–10 ft)
Zone I	A structureless reddish brown clay sand or sandy clay often divisible into Sub-zone Ia an upper non-mottled layer, and Sub-zone 1b a lower mottled horizon	*c.* 10 m (30 ft)
Zone II	A generally pale silty sand with the structures of the parent granite preserved and containing a subordinate amount of unweathered rock as corestones. May be sub-divided: Sub-zone IIa containing less than 10% corestones Sub-zone IIb containing 10–50% corestones	*c.* 30 m (100 ft)
Zone III	Contains dominant corestones (over 50% or other unweathered rock fragments)	*c.* 8 m (25 ft)
Zone IV	Massive, slightly weathered, solid rock with less than 10% debris.	*c.* 8 m (25 ft)
Fresh Bedrock		

(After Ruxton and Berry 1961a)

The interface between the weathered and fresh bedrock was described as the *basal surface of weathering* and is of particular importance to the theme of this study. In the granite profile it is usually shown at the base of Zone III rather than Zone IV (see figure 19, p. 85). Mabbutt (1961b) has stressed that this is an active *weathering front* rather than a static surface.

Variations from this description have been noted by Ollier (1969), and are:

(1) The recognition of a definite lateritic profile containing a mottled and pallid zone (Mabbutt 1961a, and figure 12, p. 54).
(2) The specific identification of exfoliation or sheeting plates, above the fresh rock.

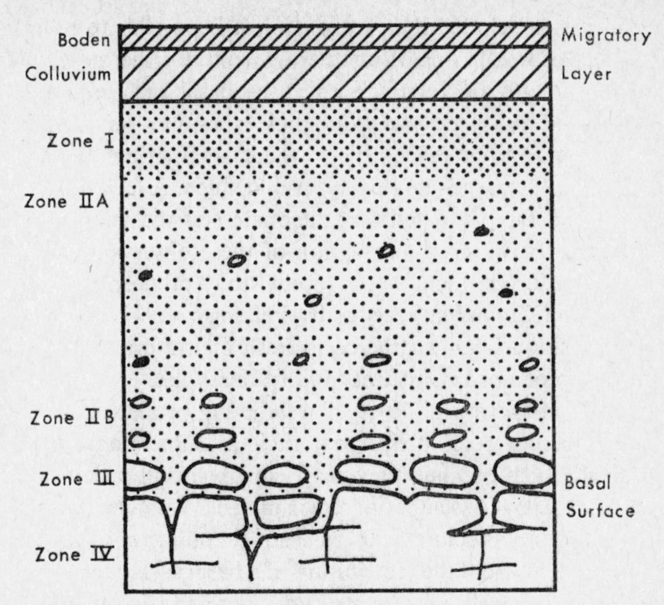

Figure 19 Characteristic weathering profile over jointed granite (after Ruxton and Berry 1961a).

For description of zones see table 9.

(3) The presence of a zone of locked corestones, intermediate between Zones III and IV, and within which decomposition is confined to joints, and spheroidal rounding is minimal.

Barnes (1961, p. 48) gives a clear description of a typical weathering profile from the Basement Complex gneisses of Uganda: '(a) soil, (b) laterite, sand or clay, (c) highly weathered and incoherent rock appearing as sand and clay, (d) moderately weathered coherent rock, (e) partially weathered to fresh rock'. The same author records that the base of (c) is seldom found below 30 m, while fresh rock is usually encountered before 50 m. This type of profile presents certain problems to the geomorphologist, for if the transition from regolith to fresh rock is gradual through several metres then there is no real basal *surface* of weathering, only a zone of transition.

However, in most recorded cases, a marked change in mechanical and refractive properties of the rock materials can be detected at a depth corresponding with the occurrence of fresh or little altered rock, and it is concluded that, in general, it is possible to refer to a basal surface of weathering or weathering front. Furthermore, where weathering profiles are relatively thin, the basal surface is usually sharply defined in a wide variety of rock types. On the other hand many drilling records reveal that, where weathering depths are great, the basal surface is not always well defined. A lack of suitable records has so far prevented any adequate statistical test of such distributions.. However, confirmation of these observations comes from several sources, including a recent study of deep weathering patterns in the Colombian Andes by Feininger (1971) who found that the transition from weathered to fresh quartz diorite could be defined within one or two metres in most boreholes.

Variations in the morphology of weathering profiles may reflect both the inherent rock properties, such as mineralogy, texture, jointing and foliation, but they may equally result from varying conditions within the weathering environment. Weathering profiles in sedimentary rocks seldom exhibit distinct zoning, and the occurrence of corestones and the recognition of a basal surface of weathering is confined to well consolidated and more or less impermeable rocks. In other cases it may be difficult to determine the depth of weathering except on criteria of coloration and chemical analysis.

It is not always easy to relate the morphological zones of a weathering profile to the degree of weathering of the rock. This is clearly illustrated by profiles containing corestones, for around each core the intensity of weathering increases concentrically, until at a distance of as little as 50 cm (Nossin 1967) granite has been found to be thoroughly decomposed. It has also been found that below a certain size (about 20–50 cm) the corestones break down into fine gravel and sand. However, a feature of Zone I debris is the rising clay fraction and increased erosional mobility (Ruxton and Berry 1961a; Ruddock 1967). In fact, although there is frequently no simple progression of the degree of decomposition from the unaltered rock towards the landsurface, there is increasing evidence to suggest that important changes in permeability, grain size and physical mobility do correspond with the principal morphological zones of weathering profiles. In fact Saunders and Fookes (1970) utilise a scale of engineering grades found to be closely related to the zonation of the typical granite weathering profile as described by Ruxton and Berry (1957).

This becomes particularly important where landsurfaces and soils are developed over truncated profiles, a situation which is common in deeply weathered terrains that have been affected by climatic and/or tectonic changes.

The development of weathering profiles

In a similar way the development of characteristic deep weathering profiles is an expression of the progress of chemical weathering. It is often difficult to say whether a given profile is active or relict; of ancient or of recent origin. There are, however, certain features which although they are not free from ambiguity offer some guidelines in the matter. It is important to realise that mere depth of weathering may be misleading in this context. Many granites, for instance, become disaggregated to considerable depths, with but minimal chemical alteration. This has been noted particularly in temperate areas such as North America (Eggler, Larson and Bradley 1969) and in New Zealand (Thomas 1974). Such profiles seldom display corestones and may be a result of particular weaknesses in the rock. Eggler *et al.* (1969) in North America attributed this feature to very early, slight alteration of the biotites, prior to weathering attack. Such deep disaggregation, exhibiting only a very low clay content and with the feldspars mostly preserved can probably be attributed to comparatively recent alteration under conditions of weathering which do not promote thorough decomposition of the rock.

In cases where corestones are frequent throughout deep profiles within which the intervening debris is thoroughly weathered to a kaolinitic, sandy clay, it can be suggested that continued weathering penetration is active. Although Ruxton and Berry (1957) considered the characteristically zoned profile to be a result of prolonged weathering, in many cases corestones appear to be forming continuously, during active contemporary decomposition. Essentially similar deep-weathering profiles over granite can be seen, for example, in New South Wales, at all levels in the landscape, from near the summit of Mount Kosciusko, at altitudes above 1300 m down to near sea level. The inference to be drawn from this situation is that the deep weathering is not to be associated with a single land surface of great age, or with conditions of extreme planation. Many corestone profiles suggest immaturity and active development, as on steep hillslopes where the cores move down the slope as the surrounding matrix of saprolite is removed, and they appear to be replaced as weathering attack continues.

Although deepening of the regolith depends on a slow rate of surface removal, it is incorrect to suggest that all deep weathering must take place beneath senile landsurfaces. Relief development itself may increase the rate of weathering by accelerating the movement of groundwater through the rock, a process which will be assisted by the tensional opening of joints within the rock mass, as dissecting streams open up deep valleys.

As with soil, the appearance of distinctively zoned profiles may be an indication of surface stability and prolonged development. Deep lateritic profiles with thick pallid zones are certainly always a result of long periods of weathering, during which surface removal has been negligible. Granite profiles in which corestones are found at depth beneath thoroughly decomposed material are also presumably much older than those where the cores are encountered near the surface. However, there are dangers in generalisation here, because corestones may persist in the upper part of the profile, and may be separated from neighbouring cores by depths of up to 10 metres. The development of these granite corestones depends upon a suitable texture, and jointing in the rock, and they will not appear where the rock is weakened throughout by some mineral alteration. They are best developed in coarse, porphyritic granites which are well jointed, and may appear in granodiorites, but are less often found in metamorphic rocks such as gneiss. In the latter the absence of a well-defined joint lattice, and the presence of foliation and banding of minerals produces different profile features. In gneissic rocks, however, an abrupt basal surface commonly transgresses the foliation of the rock, although the few isolated cores that are seen, commonly disintegrate along partings corresponding with the mineral banding, as a result of differential weathering penetration.

In summary it becomes clear that profile characteristics vary according to rock type, texture and fracture patterns, and also according to the site history. The tendency to equate all deep weathering with lateritic profiles, and hence to consider them to be of great age, is seriously misleading.

Weathering depths and patterns

Maximum depths of rock decay and the extent of deeply weathered rock are both topics of considerable importance in tropical geomorphology. But our knowledge of subsurface conditions is in reality so scanty that few generalisations are justified. Attempts to relate maximum weathering depths to climatic parameters such as rainfall (Ruxton and Berry 1961a) or to warmth of climate have been singularly unconvincing (Leopold, Wolman and Miller 1964). However, the reasons for this are varied. In the latter case the figures given by the authors give a misleading idea of both distribution and the maximum depths found in any given area (Leopold, Wolman and Miller 1964, table 4.6, p. 124). It is less easy to improve on their figures however. In the depths quoted by Berry and Ruxton (1961; see p. 13) there is an assumption that present climate, and in particular moisture availability, is the major controlling factor.

However, apart from the importance of lithology, two related problems arise: first, the age of landsurfaces determines the period and therefore to some extent the depth of weathering penetration; secondly, climatic changes involving both wetter and drier conditions have affected most of the tropics outside of narrow equatorial and desert zones. The periodicity of these changes is such that they must have a great influence upon all ancient landsurfaces. It is true however that Ruxton and Berry (1961a) apply their figures specifically to the footslopes of major hillslopes, and therefore by implication to essentially recent landsurfaces. It was an important part of their reasoning concerning deep weathering in Hong Kong that the phenomenon was regarded as a product of Pliocene and Quaternary time. It is not correct therefore to compare directly such measurements with those given by Ollier (1960), for instance, beneath the remnants of pre-Cenozoic surfaces in Africa.

There are many dangers inherent in giving maximum recorded depths of decomposition as a guide to conditions in an area. In some cases records may not be sufficiently widespread (or may consciously be placed where weathering depths are predicted to be shallow); in other cases freak figures of great magnitude may be extremely misleading. Thus in the Zambian Copperbelt, slight alteration of the basement rocks has been found at depths of over 1000 m (Mendelssohn 1961; Rhodesian Selection trust 1963—

personal communication). High figures are also given by Ollier (1965, 1969) from Australia where more than 120 m of alteration was found during tunnelling for the Keiwa hydro-electric project. Even if hydrothermal alteration can be discounted these instances are probably highly exceptional occurrences localised by shatter belts or by other structural features.

Widespread decomposition of igneous and metamorphic rocks on the other hand extends to 100 m in certain areas and is commonly more than 30 metres. At Fort Trinquet in the Mauritainean Sahara, where the present rainfall is 50 mm, Archambault (1960) has recorded basins of decomposition within the granite of the order of 20 metres. Mabbutt (1965) and others have recorded weathering depths of more than 30 m in central Australia, where rainfall today is in the region of 250 mm, whilst relict hills at Kano in northern Nigeria, where there is a seven month dry season and an annual rainfall of 750 mm, preserve an ancient regolith of at least 50 m in thickness. Comparable depths of decomposition are found in the granites of the Jos Plateau, Nigeria, which experiences a rainfall of 1125 millimetres. In East Africa similar figures have been obtained. Bisset (1941) found in Uganda that the weathered zone is generally from 30–45 m deep, and this appears to be a characteristic range over many of the basement rock areas of former Gondwanaland. Within basic rocks and metasediments, and also in certain sedimentary rocks, depths of decomposition may be much greater. Nagell (1962, p. 483) records from Brazil that in an area of dissected metasediments 'no rocks outcrop on the ridge crests and are only sporadically present in the beds of small streams tributary to the rivers. Most exposures are in the main river beds, while beneath the ridge crests weathering extends to depths of more than 100 m, or nearly to local base level.' Feininger (1971) found similar depths (up to 90 m) in quartz diorite in Colombia. Similar figures of weathering depths can be quoted from many other areas not all of which lie within the tropics (see chapter 11).

It appears therefore that weathering may extend to great depths, but only locally exceeding 150 m, and then probably due to special circumstances. Elsewhere in the humid tropics it is probably unusual to find maximum local depths of less than 20 m, even in areas of pronounced stripping or relief development. Büdel (1957) gives 30 m as an average figure and 60 m as the extreme. Similar depths to this have been recorded from Cretaceous sedimentaries in Nigeria (De Swardt and Casey 1963).

Under normal circumstances therefore it would appear that there are certain factors limiting the penetration of groundwater and the deep decomposition of rocks. One such factor is of course the surface removal of regolith, another is the possibility that confining pressures in rocks keep potential joints closed to groundwater circulation below depths of approximately 100 metres.

Spatial patterns of weathering

There are surprisingly few records of the patterns of deep weathering in relation to the configuration of the landsurfaces and much of the following section comes from my own collation of records from Nigeria (see Thomas 1966a).

The sites discussed in this chapter are spread over some 4° of latitude, embracing the humid forest country in the south, where mean annual rainfall exceeds 1500 mm and the dry season lasts barely 4 months, and also the sub-humid, savanna lands of northern Nigeria, where rainfall amounts may be less than 850 mm and the dry season as long as 7 months.

No attempt will be made at this stage to consider the possible significance of such climatic variations to the development of deep weathering profiles, because little evidence can be adduced from the information obtained from the profiles. Furthermore, the important known Pleistocene variations in climate largely obscure differences in the current rates of deep weathering between one place and another. Variations in susceptibility of the rocks forms another factor in this complex equation that we are not in a position to solve without much additional knowledge.

Study of a number of sites in Nigeria (figures 20–24) permits a few general observations. In the first place marked irregularity of the basal surface of weathering is characteristic of most of the profiles. This irregularity can be shown to consist of a series of basins and domical rises in the rock surface both in the area near Jos (figure 20), over biotite granite, and also in basement gneisses around the Owu River in western Nigeria (figure 22). In the former case a correspondence exists between the orientation of the basins of weathering and the dominant joint directions in the area (figure 21). On the other hand, transverse profiles across the Oshun and Oba

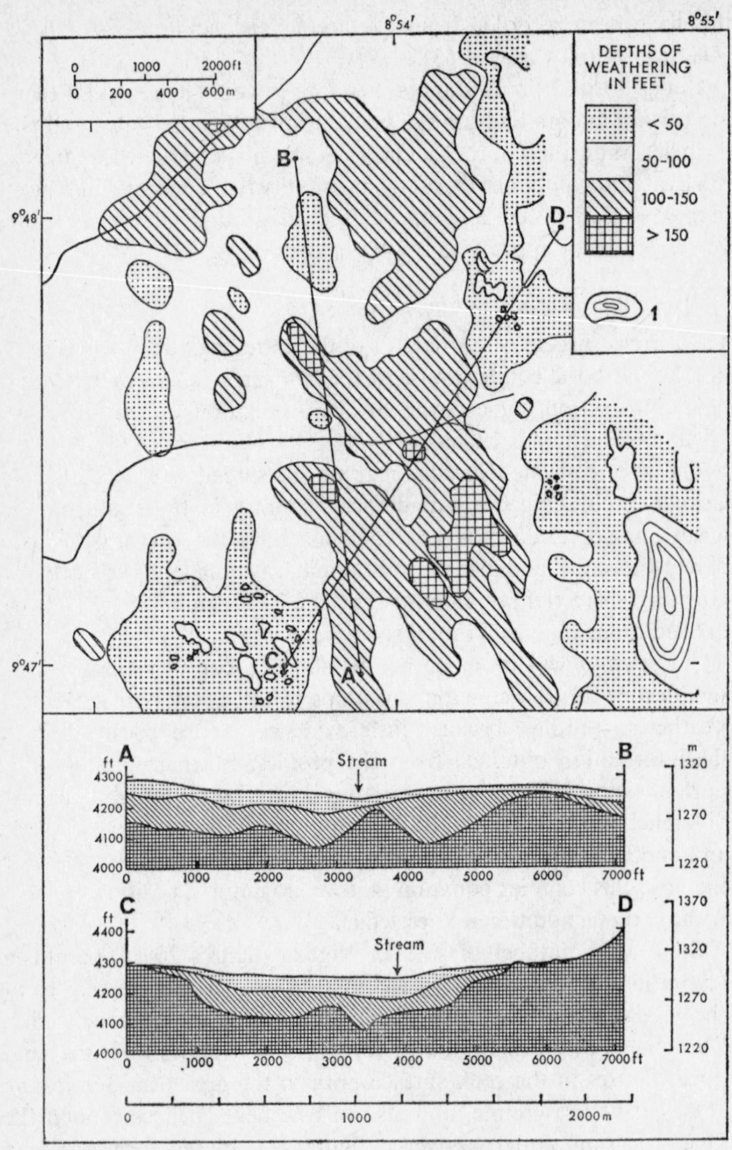

Figure 20 Deep weathering patterns in fine-grained, biotite granite near Jos, northern Nigeria.

1. Outcrops of unweathered granite, showing contours at 50 feet (16 m) intervals. On the cross-sections (A–B; C–D) sedentary regolith is diagonally ruled; the stippled material represents an alluvial fill.

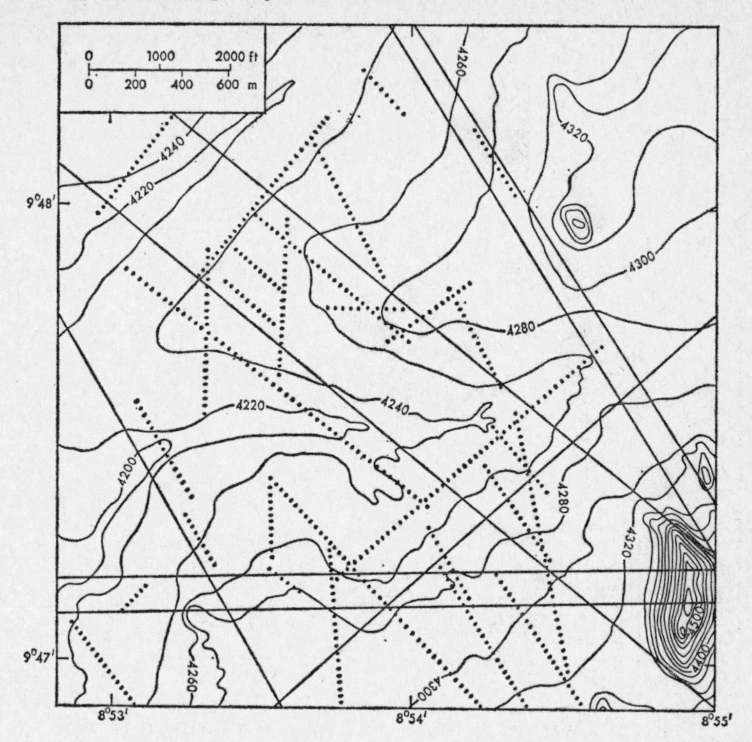

Figure 21 Topography and jointing in the area shown in figure 20.

The firm lines show projected joint directions from observed lineations on outcropping granite; the dotted line indicates corresponding lineations in mapped weathering troughs (figure 20).

Rivers (figures 23 and 24), and also weathering on the right bank of the River Niger (Thomas 1966a), suggest that an irregular increase of weathering depth with distance from the river channel may be a common occurrence; if so it must have some general geomorphological significance. This observation has been corroborated by Ruddock (1967) working in southern Ghana, and more recently by Feininger (1971) from Colombia. Similar features were also shown by Ruxton and Berry (1957) from Hong Kong.

The maps and profiles drawn by Feininger (1971) show a striking

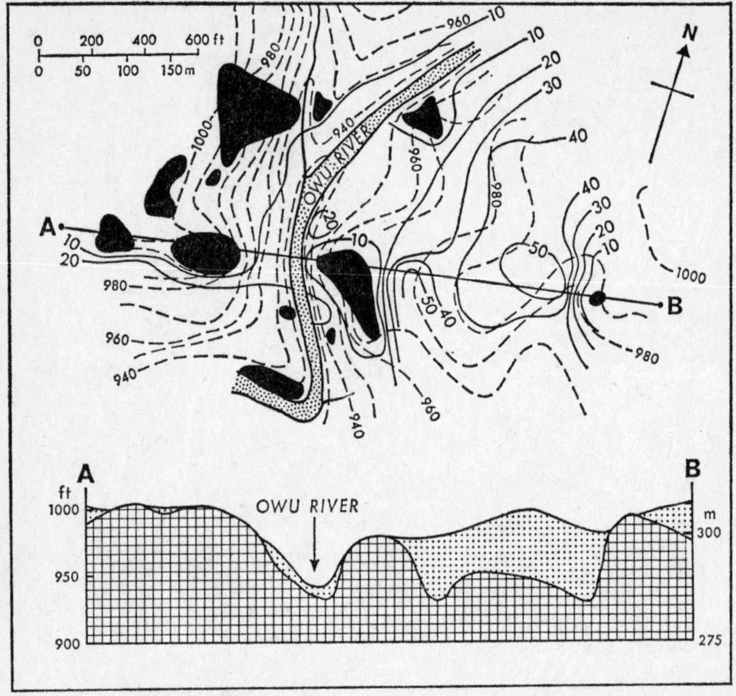

Figure 22 Weathering depths and outcrop patterns over Precambrian gneiss around the Owu River, Nigeria.

Rock outcrops shown black.

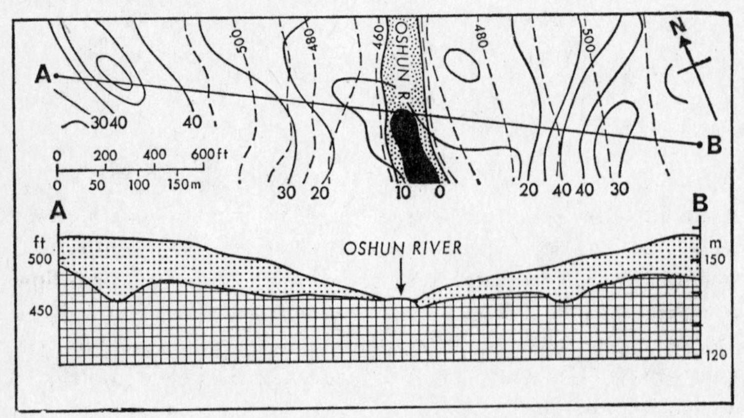

Figure 23 Weathering depth in relation to valley form over Precambrian gneiss across the Oshun Valley, Nigeria.

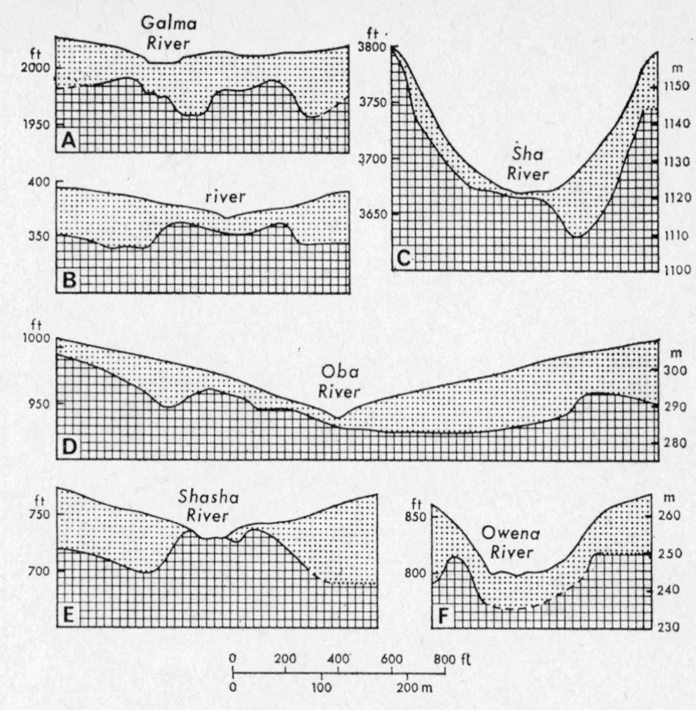

Figure 24 A selection of transverse valley profiles from Nigeria showing contrasting weathering patterns.

A. Galma River near Zaria (11°N), Nigeria (Precambrian quartz-schist); B. Igbetti Dam site near Ilorin (9°N), Nigeria (Cambrian, Older Granite); C. Sha River on the Jos Plateau (9°30′N), Nigeria (Jurassic biotite granite); D. Oba River near Ogbomosho (8°30′N), Nigeria (Precambrian gneiss); E. Shasha River near Ife (7°30′N), Nigeria (Precambrian gneiss); F. Owena River near Ife (7°30′N), Nigeria (Precambrian gneiss).

resemblance to those derived from records in Nigeria, and he comments particularly on the progressive thinning of regolith from summit locations towards valley bottoms, where most streams flow over sound rock. This occurs 'remarkably uniformly in areas of moderate relief (around 100 m), but becomes reversed where very high relief is found. In the latter areas colluvial valley fills may be combined with rock exposures on upper slopes'. Feininger found that the basal rock surface exhibits 43–78 per cent of the relief found on the upper topographic surface. Some of his figures are shown here.

TABLE 10

AVERAGE VERTICAL THICKNESS OF THE WEATHERED ROCK MANTLE AT THE SANTA RITA AND PUERTO VELO SITES, EAST OF MEDELLIN, COLOMBIA

	Santa Rita site			*Puerto Velo site*	
Bedrock	Quartz diorite hornblende diorite			Quartz diorite	
Mean elevation	1890 m			750 m	
Local relief	125 m			135 m	
Mean annual temperature	17°C			27°C	
Mean annual rainfall (est.)	4950 mm			2400 mm	
Topographic location	Average vertical thickness (m)	Standard deviation (m)	Number of borings	Average vertical thickness	Number of borings
Hilltop and ridge	46	13·5	28	45	9
Upper half of slope	36	8·0	27	43	1
Lower half of slope	26	8·0	47	31	2
Valley bottom	17	7·5	27	10	1
All	31	14·0	129	40	13

(After Feininger 1971)

An increasing number of new road cuts reveal the marked undulations of the basal surface, and the frequency of deep interfluve weathering, even in the absence of a protective duricrust capping is clearly seen from widely separated locations (plate 6).

However, interfluve weathering appears limited to areas of low or moderate relief in humid areas with a natural cover of evergreen forest. Exceptions arise where the profiles are protected by thick duricrusts, but elsewhere in regions subject to strong surface erosion across poorly vegetated slopes, deep profiles appear restricted to valley floors and other topographic lows (Mabbutt 1961a).

Other profiles illustrated here indicate weathering beneath the stream channel itself (figure 24). and it may be possible to suggest that the rivers become 'superimposed' upon the irregular basal surface of weathering if and when incision into channel deposits occurs (Thomas 1966a and figure 43). Although weathering beneath the stream channel appears confined mainly to small seasonal streams in areas of low relief, weathering undoubtedly penetrates below the *level* of the stream channel beneath valley sides, even where the stream itself flows over fresh rock (figure 24E). This was also observed by Feininger (1971) who points out that the valley forms are not reflected in the fresh rock surface. He also describes mule tracks that have been worn down in the weathered rock to as much as 3 m below the level of adjacent streams.

The features displayed by the deep weathering patterns near Jos itself (figure 20) are of considerable interest. The map shows the depths to unaltered bedrock within a small area due south of Jos on the main plateau surface at around 1400 metres. It lies within the Jos–Bukuru Ring Complex, and the weathering is within a closely jointed and fine-grained biotite granite. The weathering pattern displays a series of discrete basins and domical rises in the basal surface of weathering which breaks the surface to form outcrops in places. There is a striking conformity between these patterns and the joint sets mapped from aerial photographs (figure 21).

These patterns agree with the observation made by Enslin (1961, p. 379) in South Africa, that the weathered rock 'tends to vary in depth and lateral extent, and to form a large number of isolated, basin-shaped or trough-like groundwater compartments. These areas show practically no surface indications, but can be differentiated by geophysical methods.' Feininger (1971) also corroborates the existence of such patterns.

Around Jos as in many other areas, the sections show considerable accumulations of superficial sediment. The area has clearly undergone periods of stream incision and alluviation associated with strong hillslope erosion. Fundamentally controlled by oscillations of climate and vegetation cover, the pattern is also much disturbed by vulcanism which probably persisted into the Pleistocene here as in the East African Rift Valleys.

In this area too a platform of shallow weathering sometimes extends for a short distance around outcrops, and then plunges steeply into the deep basins of decomposition (figure 20, Section C-D). In some cases the rock surface dips steeply beneath the regolith, without any sign of such a platform (figure 22).

It is difficult to offer generalisations concerning weathering patterns because we have too few records to establish clear relationships.

However, it is interesting to summarise available observations, in the hope that they may guide future enquiry.

Summary of observations on deep weathering patterns

(1) The form of the basal surface of weathering may be highly irregular and often exhibits a pattern of discrete basins and domical rises.

(2) Such weathering patterns appear to be aligned with the fracture system observable in surrounding rock outcrops, though this does not permit any predictions as to the depth.

(3) Weathering beneath stream channels is restricted in the main to small, seasonal streams with gentle gradients, and flowing over senile plains; in other cases the streams flow predominantly over sound rock or over alluvium.

(4) In the humid tropics the depth of weathering often increases irregularly away from streams and towards adjacent interfluves. This may represent a general relationship wherever the landscape has reached a certain stage of development, and where marked inequalities in the resistance of the rocks are absent. Wherever massive rock occurs it is likely to form a rise in the basal surface which may break the landsurface to form domical outcrops of varying size.

(5) In the drier savannas and semi-arid areas observations suggest a reversal of this tendency (4), with deep profiles being restricted to depressions in the landscape, except where they are protected by duricrusted hills (Mabbutt, 1961a, 1965). High relief and steep slopes may contribute to a similar pattern of rocky hills and weathered valley floors in both humid and arid areas.

(6) In some instances deeply decomposed rock is contiguous with outcrops, especially of granitic rocks, but this is certainly not always the case. Intervening surfaces may exhibit only shallow weathering, or benches of shallow weathering may surround the outcrops, particularly where well-jointed rock occurs.

In fact patterns of surface relief offer few clues to deep weathering patterns and many of the observations made here need rigorous testing.

4 Denudation in Tropical Environments

In the previous chapter discussion centred upon the *in situ* alteration of rock materials under tropical conditions. However, movement and translocation of dissolved ions forms a major part of such an analysis, and the transported elements commonly become immobilised at new sites which may be geographically and topographically separate from their source locations. Thus there is clearly only a very indistinct dividing line between *in situ* weathering and the removal of weathered products which may be regarded as denudation. The most mobile elements in the weathering equations remain in solution, and it has been established that removal of these solutes is necessary if reactions leading to continued rock decay can continue. Ultimately such groundwater movement must be lateral and downslope, eventually seeping into streams which carry the water and dissolved salts into the ocean.

This 'chemical denudation' must therefore be regarded as a natural and important corollary of chemical weathering. It is unfortunately a process which is difficult to evaluate relative to the total effects of all denudational processes. For although estimates of dissolved solids in river water abound, it is difficult to reason from them to measures of landform lowering. First, although ions are lost in solution, residual materials become hydrated and expand in volume, thus compensating for the loss of other compounds. This is accompanied by an increase in the voids within the weathered material, so that a simple translation from the chemical constituents. of river water to a statement concerning denudation is not easy. Nevertheless there must be a denudational effect from such processes and some discussion of the topic is necessary in this context. It is also a matter related intimately to the questions concerning laterite formation raised by Trendall (1962), see p. 67, chapter 2.

Closely related to the removal of material in solution is the question of mechanical eluviation of fines from within the profile. The slow movement of groundwater through the regolith (throughflow) may carry with it important amounts of silt and clay particles; some of the latter in colloidal state. This process of subsurface mechanical eluviation partially accounts for the increased porosity of many weathered rocks and has other and important geomorphic effects as will be demonstrated later. This process is largely confined to heterogeneous regoliths such as those produced from granites, which allow the finer particles to move from between the larger ones. In addition, this process probably adds to the suspended sediment load of many tropical streams although, as will be discussed later, this is not as high as was formerly predicted.

Regoliths derived from basic rocks, especially in humid tropical environments, are of more uniformly small particle size. Eluviation is therefore less effective and the passage of water through the regolith may be very slow, although porosity is high. Under these conditions overloading, and mass movement of regolith become very important. Mass movements of a more gradual kind may be widespread throughout the humid tropics. The effects of surface wash on tropical regoliths are also difficult to evaluate. High permeability (or infiltration capacity) is a feature of many such regoliths and runoff on moderate slopes ($<15°$) beneath a forest canopy may be minimal. However, it is perhaps in this context that climate and vegetation play their most important role in distinguishing the forest denudation systems from those of the savannas and steppes, where the erosive effects of intense rainfall on bare soil materials are most potent.

Lineal erosion by streams in tropical environments is commonly regarded as feeble, on account of the fine grade size of the weathered material forming the load of the streams, and the fact that the materials show an important mobility break between the boulder and fine gravel size fractions. With a high clay content and a low pebble and coarse gravel load, such rivers are often considered to be unable to induce vigorous valley development. Absence of significant dry season flow in many rivers also diminishes their erosive capacity. Each of these processes may be discussed individually, although it is their combined effects which produce particular landform landscapes.

Chemical denudation

It might be anticipated that within environments where chemical weathering is rapid and thorough, the removal of material in solution by rivers would also be particularly effective. But the indications of much recent work suggest that this has commonly been overestimated. There remain, however, a number of ambiguities in the equation. Studies of the mineral content of tropical river waters have shown that the amount of dissolved silica, although it may be higher, is not very much greater than for extra-tropical rivers. Davis (1964) showed from a variety of sources (though biased in favour of North American examples) that the median value for silica in groundwater was 17 parts per million (p.p.m.), with a range from 12 p.p.m. to 23 p.p.m. The median value for river water was given as 14 p.p.m. which compares with the figure of 13 p.p.m. found by Livingstone (1963).

For tropical rivers Davis (1964) suggests a higher value of 24 p.p.m., based on African examples, while Douglas (1969) examined the silica concentrations of small streams in both Queensland and Malaysia and obtained a range of figures from 10–25 p.p.m. Recently Grove (1972) found that samples from the Niger, Benue, Logone and Senegal Rivers had a range from 5–17 p.p.m. with a median value of 10 p.p.m.

It is clear from these figures and from other comparisons that there is no conclusive evidence for high silica concentrations in tropical rivers, and therefore any suggestion that a high rate of silica removal takes place in the tropics must rest upon evidence of higher stream discharges. Douglas (1969) established a significant relationship between total silica load and runoff, and it would thus appear that high rates of chemical denudation as indexed by silica removal depend largely on runoff. Corbel (1959) suggested that chemical denudation from this source rose from around 0·9 $m^3/km^2/yr^{-1}$ in temperate oceanic climates to from 4·0–9·0 $m^3/km^2/yr^{-1}$ in humid tropical climates, seasonal climates in the tropics having widely varying values. But it is unlikely that we can accept these figures without reservation or that they can easily be interpreted in morphological terms.

Some interesting questions arise from the apparently low concentrations of silica in tropical ground and river waters. Under laboratory conditions the solubility of silica rises by 50 per cent if the temperature is increased from around 5°C to 25–30°C, at which level the equilibrium concentration is around 135 p.p.m. However, much of the silica released by weathering processes is either cycled or fixed within the weathering system. Lovering (1959) has shown that large amounts of silica are taken up by plants, especially the deep rooting trees of the tropical forest, and these amounts are cycled within the system, as leaf litter accumulates on the soil surface and the silica is leached back into the soil. On the other hand much of the silica is fixed within relatively stable clay minerals, and Davis (1964) suggests that the presence of large amounts of segregated iron and aluminium oxides greatly reduces silica solubility and favours immobilisation as kaolin clays. Nye and Greenland (1960), Moss (1969) and Douglas (1969) have all emphasised the nearly closed system represented by the tropical rain forest, in which uptake of nutrients commonly exceeds losses in drainage water for most common minerals including also potassium, calcium and magnesium. Such factors possibly help to explain why silica concentrations in groundwater may be much higher than those in river water. Trendall (1962), for instance, found concentrations in the region of 50 p.p.m. in borehole water in Uganda. From the foregoing remarks it may be predicted that much of the dissolved silica never reaches open stream channels, and that that which does becomes very much diluted.

Another aspect of this problem recently observed by van Schuylenborgh (1971) is the generally lower silica content of tropical (red) soils than temperate (brown) soils, in spite of similar concentrations in stream water. This is attributed by van Schuylenborgh to the greater age of most tropical red soils and regoliths. This factor may be of very great significance to this and many other arguments concerning tropical weathering and landform development, since so much of the tropical zone is comprised of old and relatively stable land masses.

When the total dissolved solids in river water are considered, the same relationships to total runoff are seen (Livingstone 1963) and tropical rivers are not found to have particularly high concentrations. Douglas (1969) also found that the suspended load in rivers responded in a similar manner to variations in runoff, but here the degree of variation was much greater from time to time in individual catchments. This will be discussed in a subsequent section.

It is difficult to evaluate the effects of these processes upon rates of denudation, for although the figures suggest that the tropical zone is little different from the temperate areas in terms of removal of solutes by river water, in reality the situation may be more complex. The varying amounts of runoff in different climatic and vegetational zones are clearly important, and the drainage density will also influence the total denudational effect of such processes. It is also possible to express the proportion of dissolved salts to suspended sediment as a percentage. Corbel (1957, 1959) thus calculated that this proportion rises from 34 per cent in the seasonal tropics to 70 per cent in the rain forest. These figures are corroborated by Rougerie (1960) who gives 60–80 per cent as the range for large rivers in the Ivory Coast. Over a small granite catchment this figure rose to 95 per cent.

On the other hand, figures from large catchments in South America come closer to the lower figure quoted by Corbel. Thus Gibbs (1967) gave the proportion of total erosion achieved by solution for the Amazon as 32 per cent, and Depetris and Griffin (1968) quote 35 per cent for the Rio de la Plata. In both these studies the authors stress the importance of the mountainous areas within the catchment to the production of both dissolved and suspended sediment. Such figures make it necessary to view with caution generalised rates of erosion for tropical regions when these are based on figures from large heterogeneous catchments (Holeman 1968).

Within the landscape the denudational effect of solution may be highly variable. High pH values for water in contact with bare rock surfaces may indicate that removal of silica from such surfaces is important. The studies of Ruxton (1958) concerning inselberg hillslopes in the Sudan also indicate the importance of the solutional component in denudation beneath a shallow weathering profile. These calculations were based not on the chemical content of groundwater, but on relative proportions of quartz, feldspar and residual silt and clay in pit samples. These figures were then related to the quartz : feldspar ratio in the original rock to give net losses by solution and mechanical eluviation. Where alteration had proceeded farthest with the elimination of all feldspar, he found a total loss of 75 per cent roughly half of which (38 per cent) was due to solutional removal; the remainder was due to eluviation. Mean figures amongst samples for sedentary material were 34 per cent removed by solution; 27 per cent by eluviation, and for migratory samples 32·5 per cent and 24·5 per cent respectively. These figures suggest that more than 50 per cent of the rock was being lost by these two processes.

However, this was not a volumetric analysis and it is therefore again difficult to translate into denudational terms. Hydration of residual materials may well make up for a high proportion of the calculated losses. Some further problems related to the input of dissolved salts through precipitation have been reviewed by Goudie (1970).

Mechanical eluviation

The selective removal of fine grained material in the silt and clay fractions from regolith mantles by the slow movement of groundwater has been shown to comprise an important process in parts of the tropics (Ruxton 1958). Catenary relationships commonly exist on hillslopes in these regions and may display the effects of this process whereby fine material has been removed from the upper slopes but accumulates in the lower members of the catena. Such a process increases the permeability of upper slope soils, but decreases that of the lower slope materials. Some of the fine material may be transported out of the profile to contribute to the suspended load of streams or in arid areas accumulates with wash material to form clay plains. It was indicated by Lumb (1965) that leaching or eluviation of fines may account for the differences between theoretical and observed porosity and clay content in tropical regoliths, a point taken up also by Ruddock (1967).

According to the account of Ruxton (1958) this process is rapid on steep hillslopes, where the groundwater moves relatively fast and is frequently renewed. Strong leaching is thus accompanied by eluviation, and this is probably a general relationship. It is also one which will be favoured by frequent rainfall, although intense falls will favour surface runoff, and also by regoliths with marked discontinuities in grade size. Similar climatic conditions operating on regoliths of more uniform particle size may well induce different results, particularly favouring mass movement. Ruxton's interesting study (1958) also reveals certain important details related to the operation of this process within shallow weathering profiles on inselberg slopes in the Sudan. The field situation recorded by Ruxton was one where a coarse, bimodal regolith gave way progressively downslope: first to an upper pediment, below a sharp piedmont angle, characterised by moist incoherent debris, and then to a dry compacted surface deposit rich in clay on the lower pediment. The increasing clay content towards the plain, together with patches of eluviated clay on the surface of the upper pediment (figure 25),

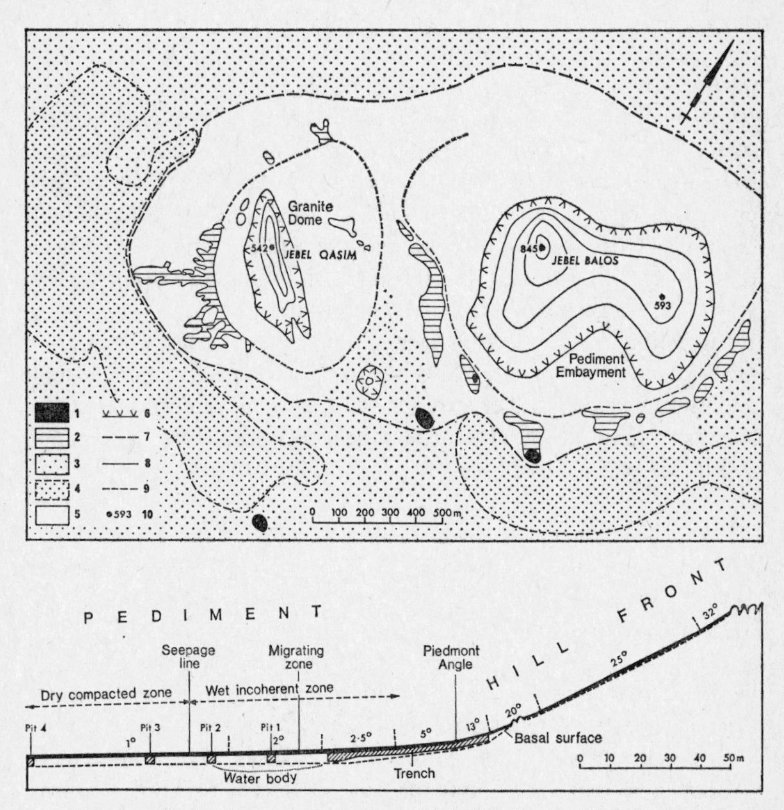

Figure 25 A. Geomorphic map of Jebel Balos and Jebel Qasim near Balos, Sudan (after Ruxton 1958).

1. Closed depressions; 2. Eluviated clay; 3. Long grass/thin bush; 4. Tall grass/thick bush; 5. Short grass/some trees; 6. Piedmont angle; 7. Sand/clay boundary; 8. Form lines; 9. Seepage line; 10. Heights: in feet a.s.l.

B. Slope profile on the south-west side of Jebel Qasim (after Ruxton 1958) Drawn at natural scale.

indicated the powerful effect of mechanical eluviation beginning on the hillslope (over 20°) and continuing beneath the upper pediment. The morphological implications of this situation are explored elsewhere (chapter 8).

Nossin (1964) considered this process to be important under humid tropical conditions in Malaysia, where removal of fines from granite regoliths was held responsible for the settling and downslope movement of granite corestones. Downing (1968) has drawn attention to the formation of tunnels in parts of Natal where subsurface throughflow exploits cracks between columnar peds in shrinking clays. This throughflow removes the finer particles from the ped surfaces, leading to a widening of the cracks into tunnels. Cavitation sometimes follows this process leading to extension of dongas and other surface valley forms. Such a process is clearly dependent upon the local soil conditions, and particularly upon the occurrence of cracking clays in the subsoil. This arises only under highly seasonal and generally semi-arid conditions.

Eluviation resulting from strong throughflow thus appears to arise in a variety of situations within both semi-arid and humid areas in the tropics and elsewhere. The common occurrence of stone lines (see p. 130) may also promote throughflow which according to Birot is particularly strong at around 50 cm depth.

Mass movement

Much more is known of the processes of mass movement, because their effects are more readily seen, but there has probably been an overemphasis on the more spectacular effects of landslides and slips as against more gradual creep and colluvial processes. It can readily be shown that conditions in the humid tropics favour the occurrence of landslides and slips of different kinds, but the areas within which such conditions occur are confined to zones of high relief. Most writers agree that such forms of mass movement respond amongst other factors to slopes of 35–60°. But slopes of this order are not commonly found over the vast shield areas which cover much of the intertropical zone, except on bare rock surfaces which are largely immune to such processes. Landslides are therefore of greatest importance in tectonically mobile areas of high relief. This is not to say that mass movement is absent from the shield areas altogether; Sapper (1935) accorded the process great importance in Brazil, and observations in the perhumid areas of Sierra Leone suggest that this group of processes become important over clay rich regoliths developed from basic rocks such as the gabbros of the Freetown Peninsula and the amphibolitic schists of the Sula Mountains. It has already been noted that areas of basic rocks on the shields are commonly lateritised and stand high above surrounding plains. This high relief is often accompanied by deep dissection and the generation of long steep slopes which favour landslides.

Factors favouring mass movement in the humid tropics centre on the deep weathering process which produces a thick zone of incoherent material with properties of low shear strength, high voids ratio and high clay content. The granite regoliths are permeable on account of the residual quartz forming the coarse sand fraction, but clay-rich regoliths from basic rocks although porous are but poorly permeable since they lack the two phase grading of the acid rocks with free quartz. These clays are mostly kaolinitic, but if swelling clays of the montmorillonitic type are present the potential for mass movement is increased further. Birot (1968) has also pointed out that the common occurrence of a sharp basal surface beneath the regolith promotes sliding. It may be supposed that the contact plane becomes lubricated, often by runoff from higher rock slopes, and that sliding is induced, leading to the stripping of the rock surface. Lateritic crusts present over many such regoliths may also promote mass movement, by admitting water via tension cracks, or joints, and by virtue of its weight over the clay horizons which will become plastic and capable of flowage as a result of continuous high rainfall.

Slope factors are of basic importance, and slopes of more than 30° are commonly found necessary for the development of widespread landslipping. Above 60° on the other hand little regolith will remain on the slope or can form there. The length of slope may be important, but there have been few studies to support this.

The active processes which trigger landslides include high rainfall concentrated into a wet season, when the regolith becomes saturated and overloaded with water which raises the pore-water pressure to a point where failure occurs. Earthquakes were found by Simonett (1967) in New Guinea to be the principal factor governing the distribution and volume of landslides in the Torricelli Mountains. Mass movement can also be induced on potentially unstable slopes by tree fall which may result naturally or from high winds. On steep granite slopes, gradual processes of wash and creep may also eject large corestones from the regolith which may eventually roll downslope with the dual result of forming an unstable backslope which may collapse, and also felling small trees. These processes certainly contribute to the development of mass movements of small scale. There is, however, a familiar 'which came first' problem associated with this situation. Thus, while Douglas (1969) suggests that windfelled trees induce slipping, Ruxton (1967) implies that slips may be a major cause of tree fall.

Mass movement associated with the failure of regolith beneath laterite sheets usually takes the form of slumping, generally involving rotational movement. Moss (1965, and figure 17, p. 78) in studies of these processes in southern Nigeria showed that the duricrust itself was the main element involved in the slumping, and that this resulted from the removal of weathered sandstone from beneath by eluviation and simple flushing by groundwater. However, in different circumstances the laterite may be involved in large mass movements of the rotational slump kind which originate within the regolith. This is clearly seen in the Sula Mountains of Sierra Leone (Wilson and Marmo 1958; Thomas 1969). Here (figure 18, p. 80) an undulating upland surface at around 500 m carries from 2–12 m of lateritic ironstone over deeply weathered amphibolite and related types of schist which have been altered to a stiff clay. During the wet season, from May to November, 3000 mm of rain falls with a peak in July and August which both receive 750 millimetres. The highly permeable laterite admits much of this rainfall, leading to the development of underground channels beneath the duricrust and to an increase in pore-water pressure within the regolith. Cambering and local collapse of the laterite occurs widely and the regolith periodically suffers shear failures resulting in slumping and earthflows. These mass movements are clearly seen on aerial photographs and are concentrated around the bounding escarpments of the schist zone and the headwaters of deeply incised valleys within the mountains.

Equally high and lateritised surfaces occur elsewhere in West Africa as over the Birrimian rocks in Ghana but often show no signs of such activity, the lateritic (often bauxitic) crust remaining intact and essentially flat. In the case of the Birrimian phyllites of Ghana the clay content is probably lower (Ruddock 1967) and the rainfall is markedly less (in the region of 1500 mm).

In less deeply weathered material, and usually in areas of high drainage density and deep dissection, slumping is largely replaced by slides and avalanches, together with induced soil flowage. In Hawaii, Wentworth (1943) studied the incidence of such slides in the Koolau Range at altitudes between 600–900 m where rainfall is of the order of 2000 mm, with individual storms reaching intensities of 125–250 mm per hour. Wentworth found that over weathered basalt flows, slopes tended to be uniform within close limits from 42–48° over much of the valley side and that it was on such slopes that 80 per cent of the slides had occurred. He concluded that creep and colluvial wash maintained such slopes as stable forms except after high rainfall, when the impermeable regolith became overloaded and unstable, resulting in shallow slides or debris avalanches (after Sharpe 1938). Such avalanches were seldom more than 1 m in thickness and produced small talus fans at their foot. It appeared to Wentworth that these slides were a major cause of the knife edge character of the saddles along the main divides of the range. These movements, taking place along a frictional surface parallel to the topography and only 0·5–1·0 m thick, are quite distinct from the sub-laterite slumping.

It is possible that a parallel may be drawn between these slides and those which occur within shallow regoliths over granite and other crystalline rocks in the tropics. Such slides have been referred to by Birot (1958) and I myself have observed them in the Gbenge Hills (granite) and on the Freetown Peninsula in Sierra Leone (see plate 7), but the thickness of material involved may be much greater than that recorded for Hawaii. The similarity lies in the presence of a distinct planar surface (in fact convex in these latter instances). A recent study near Rio de Janeiro (de Meis and da Silva 1968) has emphasised the importance of joint faces in the localisation of landslides, and equally the overriding importance of concentrated rainfall in triggering actual movement. Several hundred slides were recorded in the Rio area in January 1966 as a result of exceptionally heavy rain (472 mm in 72 h), and most of these occurred in situations where the jointing of the rock, together with the presence of a weathered layer, favoured downslope movement. Alteration along curved (sheeting) joints and rectilinear fractures promoted the downslope movement of the intervening blocks. However, this study also emphasised the role of man in producing an unstable situation in this area, as a result of widespread deforestation and excavations undermining steep slopes. Evidence of much older slides coming from boulder accumulations in stream valleys were interpreted as having occurred during climatic oscillations of the Quaternary. Discussion of this paper by Usselmann (1968) corroborates these findings from the Venezuelan Andes where he concludes that much mass movement was induced during particularly humid phases of the Quaternary. But this writer contests the impeding influence of the forest in the development of certain kinds of mass movement. He refers to the occurrence of 'coups de cuillère' (slumps) under forest in Columbia.

In a study of more than 700 mass movements which occurred in Hong Kong following heavy rain of exceptional duration and amount in June 1966, So (1971) found that nearly 35 per cent of all slips occurred under woodland, although this occupied only 8·4 per cent of the land area. On the other hand fewer than 17 per cent of the movements took place under grassland, which occupies nearly 42 per cent of the area. This high incidence of mass movement under woodland vegetation tends to contradict some arguments concerning the protective role of the forest in the humid tropics (Freise 1938). The incidence of movements was, however, correlated with steepness of slope: 40 per cent of the falls occurring on slopes above 45 degrees, as a result of lubrication along joint faces, or more often as a result of the plastic and liquid limits of the regolith being reached after heavy rainfall. It seems likely that both factors are involved, but the result is in any case the localised stripping of the bedrock, a process which may be important in developing and maintaining bare rock forms in the humid tropics.

The importance of earthquake shocks has been established for a part of New Guinea where Simonett (1967) found that both the volume and number of slides varied with the log of the distance from the epicentre of a known earthquake. Most of the movements were debris avalanches, but some rotational slumps occurred on lower slopes of sedimentary escarpments. The denudational effect of these slides was calculated by Simonett to amount to the lowering of the landsurface by 10 cm every 70–100 years in the shock zone of the Torricelli Mountains (granite) and to 10 cm every 440 years outside of the seismic zone in the Bewani Mountains.

It emerges from these examples that many observations concerning the importance of large scale mass movements in the humid tropics have been made in areas influenced either by seismicity or by man. Many other spectacular instances of mass movement are clearly related to meteorological events of abnormal severity but infrequent occurrence. Wolman and Miller (1960) have warned that the rarity of such events may diminish their overall contribution to denudation rates, and that frequent events of lesser magnitude may in fact achieve more work in the landscape. However, with mass movement critical shear stresses must be reached, and in some cases at least such thresholds may only be exceeded during exceptionally prolonged and intense rainfalls such as those instanced.

Nossin (1964) in his study of Kuantan considered wholesale movement less important than the settling and creep produced by eluviation of fines from the granite regolith. On the other hand observations in the Owen Stanley Mountains of Papua shows that slump scars and slides occur widely under an unbroken canopy of rain forest on slopes of about 30° over weathered schist. Many such movements appear to take place slowly without sudden tree falls, and although they can be seen quite clearly in the slope morphology and in the detail of tree growth, they are easily missed. Many of the slumps are only 10–20 m in width and displacement, while still smaller terrace features are induced by the exposure of lateral tree roots, and local tree fall. On such steep slopes all geomorphic processes may be rapid, including surface wash (see p. 114), and the resultant forms may reflect the operation of wash, eluviation, discrete mass movements and surface creep (plate 9).

The possible influence of climatic changes on the occurrence of mass movements must also be borne in mind especially in mountainous areas. Cones of dejection containing very coarse material are found at 1600 m in the Owen Stanley Mountains, and Walker (1970) has shown that important changes in vegetation cover have occurred at 2000 m in New Guinea during the last 20 000 years. Birot (1958) attributed fossil cones and mudflows found between stripped granite domes in the Rio de Janeiro region to the effects of an arid phase of the Pleistocene. Disturbance of forest ecosystems certainly appears likely to give rise to accentuated mass movement on steep slopes. However, increased rainfall which will encourage forest growth may also induce more frequent mass movements, particularly if the pattern of precipitation involves storms of high intensity, occurring more frequently.

It was also in this area of Brazil that Freise (1938) proposed his cyclic theory, involving a spontaneous degeneration of vegetation following depletion of soil nutrients by the forest itself. This cycle involved the stripping of rock surfaces during periods of forest degeneration, presumably by wash processes as much as by mass movement. But the reason for degeneration of the forest is not made clear and authors such as Bakker (1960, 1964) and Birot (1960) have been sceptical of the efficacy of such a spontaneous process and again refer any such development to periods of climatic change (Bakker and Levelt 1964).

Where slopes are less than 20–25°, mass movement is confined to creep processes, but opinion differs concerning the importance and role of creep in humid tropical geomorphology. Soil scientists such as Nye (1954, 1955) lay considerable emphasis on the process of creep in soil formation and have divided the soil profiles into sedentary (S) and creep (Cr) horizons. At the base of the migratory or (Cr) horizon there is commonly a stone line which is comprised of angular fragments of vein quartz, but which may contain ferruginous materials and rolled gravels. The formation of these stone layers remains controversial, although there is some agreement with the view that the overlying, fine, soil must be of broadly colluvial origin. On sloping surfaces strong soil creep could be responsible for the breakdown and redistribution of quartz fragments from unweathered veins in the regolith. But the occurrence of stone lines beneath gentle surfaces and on summit locations suggests that other agencies must be responsible, at least in these cases. The occurrence of rolled fragments (D'Hoore 1964) has led to the suggestion that the stone lines are lag gravels, resulting from former periods of strong sheet wash. Brückner (1955) even claimed to have recognised dreikanter in such deposits in southern Ghana, and suggested that they marked periods of arid climate and eolian activity. Many stone lines are buried deeply by colluvial or wash material, especially in depressions. Some of this surface layer has been attributed to termite activity and considerable doubt therefore surrounds the origins of these gravels and their significance as indicators of strong creep processes (plates 11, 12).

However, soil creep, will be clearly favoured by the periodic wetting and drying of the upper soil during the wet season which follows the pattern of frequent short, but intense storms followed by periods of high evaporation. This also suggests that creep will be more important where the wet season is of longer duration.

The prominent convexity of upper slopes in areas of low relief within the humid tropics may be another indication of the strength of creep processes. Further circumstantial evidence is provided by the accumulation of corestones in the beds of streams flowing over granite terrain. These boulders frequently accumulate to considerable depths and the stream may flow entirely underground for short stretches (plate 10). It seems unlikely that all such boulders have settled into the channel from the former profile immediately above the stream, and thus it is probable that they have come from the flanks of the valley by a combination of creep, undermining and surface movement due to wash processes, eluviation, and to more rapid mass movements.

Accurate measurements for the rate of soil creep are few. Young (1963) gave a figure of 0·5 $cm^3/cm/yr^{-1}$ for 20–25° slopes in the English Pennines, whilst Kirkby (1967) found a higher mean rate of 2·1 $cm^3/cm/yr^{-1}$ over gentler slopes (median 13°) in southern Scotland. These figures were based upon the displacement of pegs in soil pits (Young 1963), but measurements of particle displacement often give much higher individual figures. Comparative figures for the tropics have been shown by Eyles and Ho (1970) to indicate a rather higher rate of volumetric movement. They quote work by Williams demonstrating a mean rate of movement in the Northern Territory of Australia of 4·4 $cm^3/cm/yr^{-1}$, on a sandstone slope of 15°. Eyles and Ho (1970) give a figure of 12·4 $cm^3/cm/yr^{-1}$ from a humid tropical site near Kuala Lumpur. Here the slope was only 10°, but the figure comes from a single pit and, as the authors themselves stress, this cannot be regarded as typical of the humid tropical zone. Nevertheless the trend of these figures suggests that rates of soil creep may indeed be high in the humid tropics and not insignificant in seasonal and even semi-arid regions. Most measurements show that soil creep drops to zero at a depth of less than one metre.

Slope wash

Assumptions concerning the effectiveness of unconcentrated wash processes under varying tropical conditions have hitherto rested on very inadequate evidence. Measurements of runoff from artificial plots have seldom replicated natural conditions, and particularly under mature rain forest we have little quantitative information. It is anticipated that the effectiveness of surface wash will increase with annual rainfall amounts within areas where the pattern and intensity of rainfall remains essentially similar. But this increase in rate of erosion is thought to fall off rapidly once a continuous vegetation cover is established (Langbein and Schumm 1958; Schumm 1965). A part of the problem in the tropical zone relates to the history of vegetation change, and particularly to the problems of the savannas. Savanna grasslands vary internally a great deal but are not characterised by continuous cover. Yet the work of people such as Keay (1959) suggests that many such grasslands are inherited from deciduous woodlands degraded by fire and cultivation. It is therefore difficult to evaluate the variation in the effectiveness of slope wash under different types of savanna and forest vegetation. This has not prevented workers such as Fournier (1962) from predicting very high rates of erosion in the savanna zone and lower rates for the humid tropics (figure 26, and see also Stoddart 1969). The degree of relief development is a major factor in such studies and Fournier (1962) incorporates a relief factor in his equations. Both Birot (1968) and Ruxton (1967) have pointed to the effectiveness of runoff beneath tropical rain forest under conditions of high relief and steep slopes. On the other hand, in the wetter savannas and forests the tropical regoliths are notably permeable and very high infiltration rates have been recorded. This is a result of the predominance of kaolinitic (non-swelling) clays and the presence of residual quartz gravel in regoliths derived from acidic rocks. This bimodal regolith is largely responsible for the almost complete absence of runoff from gentle slopes.

The role of the forest itself is less clear, and according to some authors the tropical rain forest is less effective in protecting the soil surface than its temperate equivalents. The following reasons for the effectiveness of rainwash under tropical forest have been advanced:

(1) The occurrence of canopy openings permitting direct raindrop impact.
(2) The absence of ground flora and undergrowth, thus exposing the ground surface to such impact.
(3) The thin leaf litter-humus layer resulting from rapid decay of organic matter in the tropics.
(4) The occurrence of intense rainfall during short storms and occasionally of prolonged heavy rainfall.

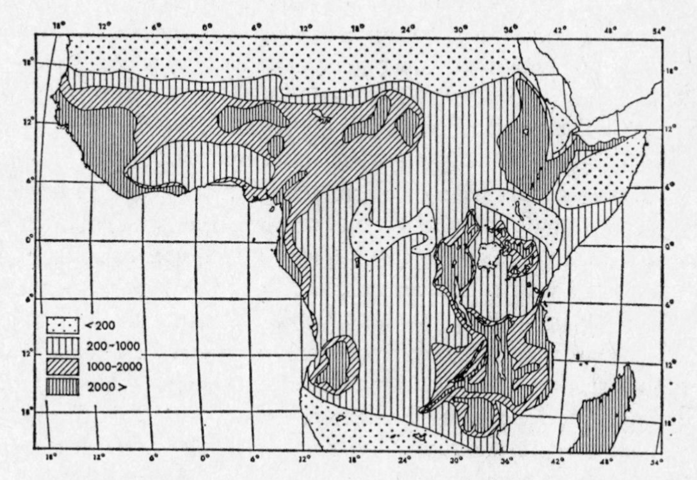

Figure 26 Rates of contemporary erosion computed for large catchments in tropical Africa (after Fournier 1962).

Figures are given in tonnes per square kilometre.

However, it is readily seen that many of these factors become important only under conditions of high relief. Thus, canopy openings resulting from tree fall are most likely on steep slopes, with shallow regoliths, where rooting depth is shallow and mass movement an important slope process. Ruxton (1967) found that leaf litter varied from 250 mm on crest slopes to less than 3 mm on the steepest slopes (35–40°).

The most important aspect of surface erosion is probably the erosion resulting from raindrop impact. Kirkby (1969) quotes random movement of 4 mm diameter stones as much as 20 cm as a result of single impacts, and considers that on sloping sites the mass downslope transport of material becomes very important, carrying particles of up to 400 mm in diameter. An important corollary of this reasoning is that the intensity of erosion from this cause will vary little across the slope, raindrop erosion being potentially equal on all sites with similar exposure. The effectiveness of raindrop erosion is very variable, and dependent upon the nature of the rainfall and of the ground cover. In seasonal climates, soil loss from raindrop erosion has been shown to decline as the wet season proceeds and the vegetation cover increases in density. Furthermore, Williams (1969) has pointed out that erosion depends on raindrop momentum (a function of drop mass and velocity) which is closely related to rainfall intensity. Early storms of the wet season in the savannas, and even under forest, are therefore likely to have a marked erosional effect. Apart from the displacement of particles downslope or randomly across nearly flat surfaces, the impact of raindrops brings about compaction of the soil surface, breaking down aggregates and sealing the pores. This reduces permeability and increases overland flow with consequential soil loss. These processes favour the development of linear slopes which are possibly maintained by the continual removal of surface water into channels, or by infiltration, so that a steady increase in runoff downslope is not experienced. The long straight slopes recorded by Ruxton (1967) in Papua accord with this reasoning. This author also noted that soils did not become progressively finer downslope, that an absence of gullies and slip scars argued for the dominance of wash processes, and that this was corroborated by the behaviour of adjacent streams which rose rapidly and carried a high silt load after heavy rainfall.

It is interesting that Ruxton (1967) concluded that slope wash processes are responsible for the maintenance and parallel retreat of constant slopes of 35–40° in this area, while Wentworth (1943) came to the conclusion that soil avalanches were responsible for comparable slope forms in Hawaii. In each case a particular process was isolated as responsible for developing a fine textured relief of steep slopes, shallow regoliths, and knife edged ridges which would inevitably be lowered by continuous slope retreat. The principal conclusion from such comparisons must be that in areas of high relief geomorphic processes act more rapidly than elsewhere and their effects are more clearly seen.

On forested sites the impact of raindrops may be secondary to the effect of tree trunks in concentrating rainfall intercepted by leaves and upper branches. This process can give rise to localised erosion at the foot of the tree and to significant overland flow extending outwards from this point. It has also been pointed out that water dripping from leaf tips will have a noticeable erosive effect, accentuated by the repetition of impact at the same spot. But whether drips from leaves commonly reach their terminal velocity before impact with the ground (about 8 m) as claimed by some writers (Young 1972) possibly remains doubtful, on account of interception at lower levels by young sapplings and shrubs.

What remains obscure, however, is the question of effectiveness of such processes under rain forest on gently sloping sites which dominate wide areas of the tropical shields. Very little mature rain forest survives beneath these sites which are favoured for cultivation, and it is possible only to point to the thick, permeable regolith usually found beneath these areas, and to the measurements given by Nye which suggest that runoff is negligible. However, raindrop impact will be effective if leaf litter is thin and where canopy openings exist. But these again will be fewer on areas of gentle relief. Soil particles which are displaced in this manner will not travel consistently downslope where this is gentle, and this again will reduce the effectiveness of runoff.

A recent study by Young (1973) has thrown some new light on this problem. Surface loss from slopes under a mature rain forest in Malaysia was monitored during a period when exceptionally high rainfall occurred. In three months, slopes over 10° lost as much as 1–2 cm of soil, presumably by wash processes. Most of this loss probably occurred during a single storm in which 300 mm of rain fell in 3 days, but similar losses were predicted for a complete year without exceptional storms. Translated into a rate of transport, Young's figures give 4 $cm^3/cm/yr^{-1}$ which is less than predicted values of creep given by Eyles and Ho (1970) as 12 $cm^3/cm/yr^{-1}$. In view of the limited duration and specific locations of these readings it is probably unwise to speculate on the relative importance of the two processes. Nevertheless it is clear that slope wash should not be discounted as an important process under rain forest, except near crests and on very gentle slopes.

Under a savanna vegetation raindrop impact will be more effective, as few tropical grasses form continuous mats of vegetation, and no

E

canopy of emergent trees can protect the ground surface. Moreover seasonal leaf shedding and burning expose large areas of soil to immediate impact from early rains which are often particularly intense falls, but of short duration. Whether such conditions have existed for very long is in itself conjectural, but even beneath an open deciduous canopy such processes might remain effective although a dense growth of rank grass would protect the soil.

An experiment by Nye (1954, 1955) carried out under a cover of Imperata Cylindrica grass, but within the forest zone near Ibadan, Nigeria, confirmed the high permeability of a deep weathering profile over gneiss. During experimental watering the plot absorbed water at the rate of 284·5 mm per hour for the first 30 min, 198 mm per hour during the next 30 min, reducing to 82·5 mm during the second hour. After 132 mm of rain had been simulated, slight seepage occurred at the stone line, but no runoff was observed and Nye concluded that lateral seepage would not be important under prevailing rainfall conditions, where individual storms do not exceed 39 mm on more than 10 per cent of occurrences.

Another indication of the effectiveness of all forms of erosion in the savannas comes from calculations of runoff from catchments in different climatic locations. Ledger (1964) has summarised much information for West African catchments and concludes with Rodier (1961) that peak flows in small streams are determined by vegetation cover. In a desert area, 60 mm rainfall produces four times the flood of an equivalent fall in the rain forest. Floods also respond more to intensity of fall in arid areas, but to magnitude in forest areas. Individual storms may result in 80 per cent runoff in areas receiving less than 750 mm annual rainfall, if relief is moderate, and annual coefficients of runoff may reach 30 per cent. However, in West Africa most large rivers have catchments with very low mean slope values and this results in channel degradation and high losses of water through evaporation, such that runoff may be less than 6 per cent per annum. In Ledger's summary (1964) map of hydrological regions the runoff coefficients rise to 25–35 per cent in the wetter savanna zones, but are nearly matched by the figures for the forest itself. These amounts decrease rapidly with increasing aridity.

The significance of these figures in the present context must be carefully considered. Runoff reaching the principal streams is not all overland flow, but the rapid response of small streams in more arid areas to individual storms suggests that much water is contributed in this manner. Sustained high rainfall in the wetter areas produces a higher runoff considered annually, but some of this figure will be from groundwater discharge. It is perhaps possible to suggest that rapid runoff from poorly vegetated slopes in arid areas will bring about redistribution of surface deposits on a large scale, but not perhaps with high losses into permanent streams, except in areas of high relief. Local redistribution of slope mantles also occurs in the humid tropics. Slumping under laterites (Moss 1965) and on steep slopes has this effect, and the accumulation of material in flat, alluviated valley floors has received frequent comment (Swan 1970a; Eyles 1968). Young (1972) has suggested that this accumulation may account for an apparently rapid slope erosion under forest, but a low rate of regional denudation in the humid tropics (Douglas 1969), where annual stream discharges are none the less high. Whether it is possible following Fournier (1962; figure 26, above) to suggest that the maximum denudation comes in the intermediate savanna zones is perhaps difficult to decide on present evidence.

Stream erosion

Although the chemical denudation brought about by tropical streams has been stressed, the efficacy of tropical rivers in the mechanical processes of erosion has been the subject of much debate. Studies of weathering and inspection, if not measurement, of stream bedloads show that crystalline rocks typically do not yield many fragments of pebble or coarse gravel size. Core boulders below a certain size which may be in the region of 0·15–0·5 m, appear to disintegrate into sand and clay as a result of a breakdown of the rock fabric due to weathering. Much of the weathered material reaching the streams by means of creep or wash processes is already broken down into fine products, and the rivers may carry only fine gravel, sand and suspended clay particles, with an almost total absence of course bedload. Under such conditions erosion by tropical streams is relatively feeble and may become further inhibited by the nature of the stream regimes which are highly seasonal; flow dwindling almost to nil in many cases, where the dry season exceeds three months in duration (Ledger 1969).

Information on stream loads is almost entirely confined to suspended and dissolved constituents. Difficulties surrounding the measurement of suspended load have been emphasised by Douglas (1968) who found that on Davies Creek in Queensland this varied from 6–58 mg/l at identical discharges. Factors affecting these readings included whether they were taken during a rising (35 mg/l) or falling (19 mg/l) flood, and also environmental factors such as rainfall intensity, and state of the soil. However, a general relationship between suspended load and discharge is clear and is illustrated by the behaviour of the Sungai Pasir (Malaysia) which has a load of 2·1 mg/l at low flow, rising to 5788·9 mg/l at peak flow. This sediment is mainly in the fine sand to clay fraction, and is probably derived principally from stream banks, where it is removed by hydraulic action of the stream and is replaced by mass movement. In itself such material gives little indication of the power of stream erosion, but we may compare figures for tropical streams given above with that for the San Pedro (Arizona) which derives from an arid catchment and carries 100 000 mg/l during flood (Holeman 1968).

Upon such observations as these has been built up a superstructure of theory concerning tropical landform evolution which follows from the basic premise that stream erosion is ineffective by comparison with other geomorphic processes operating in the equatorial and seasonally humid tropics. Leading these ideas have been several European geographers, notably Büdel (1957), Birot (1958, 1968), Tricart (1965), and Louis (1964). Büdel claims that since the rivers flow over the deeply weathered material they are denied abrasive tools for vertical erosion. This is evident from two independent observations: first, that stream valleys in tropical areas are characteristically indistinct features, having lateral slopes no steeper than the stream gradients (1·5–3 per cent), and second, that the stream sediments display the same grading as the regolith materials. Both these observations indicate the feebleness of stream erosion, and, according to Büdel, that it is not true lineal erosion at all, but an extension of the wash processes operating with great efficiency on slopes in the seasonal tropics. He suggests that stream flow has three major effects: lateral wash erosion at high water, when the stream extends over wide areas of adjacent plains; general sheet wash, also at high water, and minor gully erosion at low water, producing superficial incision into weathered material which is obliterated at the next flood.

Breaks in the river thalwegs in the form of rapids and waterfalls are common, and due to resistant (unaltered) rocks reaching the surface where they are not rapidly affected by stream erosion, and to selective erosion along the lines of major fractures (Hurault 1967). Büdel (1957) indicates the great age of many waterfalls from evidence of distinct varietal differences between water snails at each cateract on rivers in Surinam (see Bakker 1960). It is claimed that coarse fragments coming into the river at the sites of such rapids are few and persist for only short periods in the stream flow. Büdel (1957) suggests that they migrate only a few tens of metres, but Birot (1968) gives 10 km as a more likely figure.

These ideas are substantiated by observations of the load of the Angabunga River in Papua made by Speight (1965), who found that where the river debouched from its mountainous catchment the mean diameter of bed gravel was 110 mm, but a few kilometres downstream where the river has built up a large fan, the bed is comprised almost entirely of sand.

There is little doubt that these ideas have some importance to our understanding of tropical landforms, particularly in areas surrounding the gently warped basins of tropical shields. But many observations serve to complicate, if not diminish, the power of some of these arguments. Evidence presented concerning weathering patterns showed that in most cases major streams (4th order and above) flows predominantly over little altered rock. The evidence that weathering depths increase away from stream lines would appear to indicate that interfluve erosion is not rapid in the more humid areas, and is generally less active than lineal stream erosion. Furthermore, although many stream valleys are shallow it is not common for floods to extend over wide areas at high water in areas of crystalline rock. Most wide floodplains are associated with poorly consolidated and highly weathered sedimentary rocks which often lie in saucer-like structural depressions within which the streams have very slack gradients (Ledger 1969). Nevertheless, it is certainly true that many first and second order streams in both savanna and forest zones have poorly developed channels which are little more than gutters carrying dry season seepage flow. In the rainy seaon such streams spread over the valley floor as diffuse zones of wash. But although this resembles some of the observations made by Büdel, these processes take place in well-defined valleys with quite steep bounding slopes and which may be very variable in relief (Eyles 1968).

Of course, whether or not channel erosion by streams appears effective, the sediment yield of rivers is often taken to represent rates of denudation and to provide figures for comparison from area to area. This, as has been shown, rests almost entirely upon dissolved and suspended sediment loads, and although individual figures and world wide predictions for denudation rates by this means have been offered, they pose many problems. Some of these have been reviewed by Douglas (1967 (a) and (b)) who points to the major effect which man's interference with natural vegetation has had upon sediment yields. Figures quoted from the Cameron Highlands of West Malaysia show that in one catchment where 94 per cent of the natural vegetation remained intact, sediment yield was 21·1 $m^3/km^2/yr^{-1}$, whilst in another where only 64 per cent of the area was still covered by natural forest the corresponding figure was 103·1 $m^3/km^2/yr^{-1}$. Similarly, records quoted from Java indicate an increase in sediment yield from one catchment of more than 100 per cent during a 24 year period (1912–1935) during which rapid deforestation occurred. Douglas (1967b) points out that such figures make it virtually impossible to predict for large areas rates of natural erosion based upon modern records of sediment yield. This observation is also offered as a criticism of the rates given by Fournier (1960), which are likely to have been derived from catchments much modified by man. Thus, for eastern Australia Fournier gives 80 $m^3/km^2/yr^{-1}$, but measurements from catchments little affected by man show that figures may be only 10–15 per cent of this value. By the same reasoning the map prepared for Africa (Fournier 1962; and figure 26) expresses contemporary relationships rather than rates of natural erosion.

Douglas also points to the importance of natural factors in varying rates of sediment yield. Differences between montane forest and lowland rain forest in the protection afforded to the surface soil greatly favour erosion under the former, and this, combined with steeper slopes and higher relief in many such locations, probably contributes to higher rates of natural erosion. The effects of climatic change may also be important since they lead to the persistence of relict soils and vegetation. Douglas (1967b) draws attention to the survival of vegetation from wetter climates in present-day semi-arid areas. Such vegetation does not regenerate when disturbed and such a situation promotes high erosion rates.

Many of these ideas will be developed and discussed later in the book, but it is clear that, not only are contrasts in the humidity of climate and density of vegetation cover important to such theoretical questions. But the tectonic history of the areas studied is also influential, both in terms of active seismicity and its effect on mass movement, and more widely in the degree of relief development. Wayland (1934) discussed the formation of a plain from a plain in parts of Uganda, as a result of slow and prolonged uplift. In such a situation tectonic relief does not come into being, and the processes of denudation maintain a surface of low relief. By contrast, in areas of recent rapid uplift and mountainous relief, as in New Guinea, most features of tropical relief as described by Büdel are absent. As Ruxton (1967) comments: 'mountainous selva in fact often has a fine textured relief with sharp ridge crests and deep narrow ravines, implying vigorous dissection in progress. In this respect there is resemblance to some humid temperate mountain regions such as New Zealand (Cotton 1962a), illustrating that differences between morphogenetic systems are commonly less marked in the mobile mountain belts.'

Summary

It is difficult to draw conclusions regarding the efficacy of denudational processes for the tropical zone as a whole. Furthermore, even for individual catchments, or areas, it is very difficult to derive rates of denudation that may be extrapolated either in space or time. The widespread interference with natural vegetation by man has probably continued for more than 2000 years in many areas. Therefore not only are modern rates of geomorphic processes a poor guide to long-term rates of denudation, but many features of both erosion and deposition may date from the period of human occupation This may seem obvious to anyone familiar with recent and catastrophic soil erosion, but over wide areas of moderate relief such spectacular phenomena are absent, and many anthropogenic features may appear quite natural beneath a well developed canopy of vegetation.

5 Deposition in Tropical Environments

The focus of attention in geomorphology has generally been on erosional forms, but the need to associate these forms with datable deposits on the one hand, and the evidence of widespread superficial deposition which has come from the study of soils on the other, have led to closer attention being paid to both regional and local patterns of deposition. Furthermore, the study of denudational processes demands that accumulation be studied along with removal. Many recent studies have emphasised that even gently undulating surfaces of apparently great geological age may have undergone widespread superficial modification as the result of the redistribution of weathered products. Both detailed studies of soils, particularly in the Australian semi-arid zone, and geochemical data concerning sedimentary sequences of terrestrial origin, have thrown new light on the variations in intensity of weathering and erosion, during periods of climatic and vegetational change.

Concepts of landsurface stability and instability

Two concepts of landsurface stability and instability are of central importance to this theme. Erhart (1955, 1956) distinguished between the condition of biological equilibrium or *biostasy,* and periods of natural disturbance from tectonic or climatic causes which disrupt this equilibrium to create a condition of *rhexistasy*. This was applied particularly to the tropical rain forest which by anchoring the incoherent regolith, but permitting free leaching of soluble salts, leads to the *in situ* accumulation of weathered materials but the removal or separation of mobile ions (biostasy). These salts, by lateral migration via the drainage system, contribute to the accumulation of chemical sediments such as carbonates, dolomite and flint. During periods of disturbance (rhexistasy) this filter is broken and detrital material migrates with the chemical products of weathering to form clays, sandstones and arkosic sediments. Millot (1970) suggests that with the onset of changing conditions, leading to the destruction of forest, sedimentation will follow a sequence starting with chemical (largely carbonate) deposits under the forest (*biostasy*), followed perhaps by coal resulting from the destruction of the forest; clays (mainly iron and aluminium oxides and kaolin) from the reworking of lateritic profiles; sandstones from the lower horizons of old weathering profiles and finally arkosic materials from the erosion of unaltered rock itself. This analysis was applied by Millot (1970) to the great Tertiary basins of sedimentation in West Africa, where evidence of a humid tropical period persisting into the middle Eocene is to be seen in widespread lacustrine conditions and the formation of chemical sediments. This gave way to unstable conditions in the Upper Eocene and later Tertiary time leading to the widespread stripping of earlier weathered mantles and to deposition of kaolinitic clays and sandstones of the 'Continental Terminal'.

This concept assists our understanding of the relationship between denudation and accumulation on a regional scale. At a local level the concept of the 'K-Cycle'* has been developed, particularly by Butler (1959, 1967) in Australia, to provide a framework for understanding sequences of soil formation and destruction. This cycle is defined as: 'the interval of time covering the formation by erosion and/or deposition, of a new landscape surface, the period of development of soils on that surface, and ending with the renewal of erosion and/or deposition on that surface' (p. 7). This cycle gives formal recognition to the realisation that the landsurface may experience periods of stability (biostasy) during which weathering and soil profiles develop, and phases of instability (rhexistasy) when both denudation and accumulation occur. Repetition of these events leads to a series of deposits on lower groundsurfaces and against hillslopes, and on each some degree of soil development may have occurred. A form of soil stratigraphy is thus built up, but such soil sequences have not always been found to correlate over wide areas; they tend to relate to local events which may occur at different times although within a major chronology of climatic or tectonic disturbance.

These ideas are of the greatest importance to geomorphological thinking, because they demonstrate the limitations of the concept

* (K—Khronos, Greek for time)

of the steady state in landscape development over long periods of time, and also because they link landform change closely with ecological change, and both of these with climatic oscillations. Thus the study of translocated materials and of buried soil profiles confirms the periodic, even cyclic, nature of landscape change.

In areas of integrated perennial drainage much of the deposition is within a deltaic or marine environment, so that the major areas of sedimentation are geographically separate from the zones of erosion, and may not form a part of the landmass. However, in the study of tropical geomorphology, local forms of accumulation resulting from unconcentrated sheet wash, sheet flow confined within valley systems, stream deposition, lateral eluviation and colluvial processes are all important.

Most processes of mass movement bring about only local transfer of material, so that deposition occurs on lower slopes, beneath slump scars landslides and so on, although stream activity may remove much of this material to other sites. The distance travelled by other transported materials may depend upon the nature of vegetation and climate, although with the exception of stream activity most movement of superficial material appears to be of a local character. Awareness of the widespread nature of colluvial and alluvial mantles has derived mainly from the study of soils.

It is appropriate to view these depositional materials in a wider perspective both in terms of spatial and temporal distributions. The accumulation to which attention is given in this geomorphological discussion is essentially local in occurrence and mainly Pleistocene or recent in origin. On a wider scale of enquiry, large tropical landmasses have been classified into degradational and aggradational landsurfaces (King 1962; Moss 1968) which indicate, over a lengthy geological time span, the tendency for certain areas to act as receptacles for the waste of denudation, whilst others continue to rise and to possess dominantly erosional characteristics. The pattern of occurrence of such degradational and aggradational surfaces reflects the tectonic history of the landmass, and has little to do with the tropicality of climate, except in the case of large basins of inland drainage within semi-arid zones.

Thus, on the shields of Africa and Australia, large basins of a deposition have developed internally, although some, such as the Congo Basin have recently achieved a drainage outlet to the sea.

Alternations, through time, in the nature of deposition within these basins reflects the establishment or disruption of equilibrium states in the landscape, as discussed here. Peripheral zones of accumulation are generally narrow and sporadically developed. The great delta of the Niger River is somewhat exceptional in this context. But in the mobile and highly uplifted landmasses of south-east Asia, and locally elsewhere, wide coastal plains have developed from coalescent deltaic accumulations, sometimes resembling great alluvial fans as along the southern coast of Papua–New Guinea.

Local transfer of material on hillslopes

It is, however, on the local scale that attention is given here to depositional forms, for these create important elements of the landscape within areas which on the broader scale appear as zones of degradation.

Dresch (1953) and Ruhe (1956, 1960) have both referred to the widespread mantle of transported material found over tropical landscapes. These transported mantles vary in origin and depth, but can be thought of as essentially superficial over most of the landsurface, reaching depths of 2 m commonly, but increasing to many times this depth on lower slope sites. The importance of this transported mantle is illustrated by Moss (1965) from south-western Nigeria, where he describes considerable depths of a non-mottled ferrallitic sandy clay, covering some 25 per cent of the area studied. This red, friable material containing 30–40 per cent of clay, extends from the margins of laterite breakaways across gently undulating plains exhibiting slopes of 1 degree or less. Moss suggests that this deposit is not derived solely by *in situ* weathering of underlying sandstones, but as a result of breakaway retreat, due to slumping of the duricrust, following spring sapping and washing out from the underlying horizons of the weathering profile (figure 17). It is pointed out that similar deposits have been given stratigraphic significance in the humid tropical coastlands of West Africa. The so-called 'acid sands' of southern Nigeria and the '*terre de barre*' of the 'Continental Terminal' in southern Dahomey correspond in general character with these deposits, which may commonly attain depths of 6–17 m. The high clay content and poorly sorted character of the deposits argue against stream deposition. Their position and character therefore suggest local translocation from deep weathering profiles during the process of breakaway retreat. These deposits are found within the rain forest zone, where mass movement has been shown to be active. So there must remain some doubt as to whether they mark periods of climatic and vegetational change leading to increased instability of the landsurface (rhexistasy), or are a result of continuing processes of change within the forest itself. However, the areas studied in West Africa lie close to the forest–savanna boundary and contain exposed duricrusts which are characteristic of the savannas. Moss also found coarse detrital fragments below the ferrallitic clay in places.

A comparable situation is described from Western Australia by Mulcahy (1960), Mulcahy and Hingston (1961), and Churchward (1969). Here, in a strongly seasonal climate with a winter rainfall ranging from (<375 mm–625 mm), the break up of an ancient laterite profile has produced deep yellow sands (Quailing depositional surface, Mulcahy 1960) flanking ironstone residuals. These sands are mainly quartz, plus a low clay content, suggesting perhaps that they have been associated with wash processes. They attain considerable thicknesses (20 m or more) and towards the drier margins of the area form extensive 'sand plains' which were also formerly regarded as *in situ* deposits. In a neighbouring, and slightly wetter area, Churchward (1969) found that most of the soil parent materials were colluvial sands and gravels which, significantly, derived their properties from particular zones of a relic laterite profile. He concluded that each deposit has a characteristic location in respect of the drainage pattern and the degree of incision of the old lateritised surface, and that they have been derived by colluviation from respective zones in the truncated profiles (figure 44, p. 255). These examples tend to confirm that with slight erosional modification of ancient landsurfaces, extensive if very localised translocation of material takes place which results in the accumulation of colluvial and alluvial deposits over large areas of the landscape, contributing distinctive properties to the overlying soils.

The repetition of such events in the landscape may produce several ground surfaces, recognised as buried soils which might be superposed on a given hillslope, or arranged in characteristic patterns with respect to the dominant relief features. These are designated according to the K-Cycle chronology (K_1 youngest, K_2, K_3 . . . K_n oldest). Within Australia this concept has focussed attention on phenomena which might otherwise have escaped attention, but which are very extensive in the landscape. Perhaps more important, they record changes in the stability of landsurfaces within local areas and over short time periods. Such changes may record variations in climate as they have affected the vegetation cover and intensity of erosion. The sedentary materials of relic weathering profiles are seldom able to record the rhythm of climatic change; they are rather the sum of the effects of varying climate through lengthy periods of geological time. For this reason the superficial deposits offer more clues to the past history of the landscape than the erosional features which often receive more attention from geomorphologists. Butler (1959) emphasised that the unstable phases may record periods of high transporting activity, which remove material into stream channels from the sites of weathering, and also lead to the extension of erosional surfaces. Alternatively transporting power may be lowered and deposition with the extension of accreting surfaces takes place.

Actual climatic changes vary from location to location, but Butler in a later paper (1967) points out that a 'significant' climatic change is one which brings about biological degeneration which in the context of tropical and sub-tropical landscapes is likely to result from increased aridity or greater seasonality and storminess.

As one approaches the arid margins of the tropics, deposition becomes more general and widespread, forming major features such as clay playas and a general sedimentary fill within endorheic drainage basins. In semi-arid areas the recognition of pedisediment as a superficial covering to erosional pediment slopes, and of the deeper perisediment which effectively buries the pediment surface towards its downslope end are well known. In the semi-arid interior of Queensland, Connah and Hubble (1962) describe (p. 378) 'large areas of plains covered by red silt, sandy silt and ironstone gravels believed to be stripped from the high ground. Where the mass has been base-levelled, the surface is often strewn with tightly-packed residual "gibbers" or "billy" (quartzite, probably silcrete) or laterite.'

In the highly seasonal, savanna climate of the Jos Plateau (Nigeria), accumulation has taken the form of large alluvial fans and less well-defined areas of deep silty sand. Much of this material is clearly Pleistocene in age, and many of the fans are dissected by contemporary streams. They presumably witness former, more arid periods of intensive stripping from the granite hillslopes, and local deposition from sporadic sheet and stream floods.

In more humid regions the deposits may differ, but they are not necessarily less widespread. Ruhe (1956) in particular has called attention to the repeated cutting and filling which has taken place over the landscapes of the High Ituri (Congo). In areas of relief development piedmont slopes are naturally the sites of most significant deposition, and this appears equally true of humid as well as semi arid areas.

The problem of stone lines

It is perhaps within the context of the foregoing account that the vexed problem of 'stone lines' can be reviewed. It has been pointed out that these have commonly been attributed to creep, to termite activity or to the burial of surface gravel deposits (de Villiers 1965; Vogt 1966). However, the widespread development and complex stratigraphy of many sections that include stone lines, suggest that many if not most occurrences are related to comparatively short-term variations in landsurface stability. Although the claim by Brückner (1955) that they represent former desert pavements may still be regarded as unlikely at low latitudes, the possibility that these deposits are pediment gravels has recently been emphasised by Fölster (1969) and by Burke and Durotoye (1971) from work undertaken in southern Nigeria. The possible climatic oscillations involved in the development of these profiles are discussed later (chapter 10), but the general character of these deposits may also be noted here. Burke and Durotoye (1971) describe a common sequence through weathered bedrock into an iron cemented gravel, above which occurs an uncemented younger gravel, on top of which lies a colluvial, or wash, layer of varying thickness. Fölster (1969) describes some more complex sections and also attempts to relate gravel and hill wash sequences to ferruginous (laterite) crusts and to periods of shallow dissection within the weathering profile. He summarises the evidence for a depositional origin for the stone lines:

(1) The multiple stratification of the complex.
(2) The rather sharp boundaries between different layers, unusual for pedogenic horizons.
(3) Inclusion of allocthonous elements in the stone line.

(4) Abrupt truncation of petrographic features, especially quartz veins at the stone line.
(5) Occurrence of prehistoric implements of successive industries in different layers of the complex.

Whilst Fölster (1969) believes that these deposits are derived from the older weathering profiles during phases of shallow incision and redistribution of regolith, Burke and Durotoye (1971) consider the pediment gravel to be derived more directly from the retreat of slopes around residual hills, and consequently they suggest that most of the deep weathering has occurred subsequent to pediment formation. Aspects of this problem will be discussed at a later stage, but it may be pointed out that if the stone lines and other deposits date mainly from the later phases of the Quaternary as is suggested in both papers, then there has not been sufficient time to bring about weathering profiles beneath pediment gravels of from 10–20 m in depth as is common throughout the area in question (see figure 23, p. 94).

Evidence indicating the importance of termite activity has been persuasively presented by Williams (1968) who described sections over weathered granite in the Northern Territory in Australia which closely correspond with descriptions given by Nye (1954, 1955) from the rain forest zone of southern Nigeria, in which an upper gravel free soil is underlain by a gravel layer interpreted as a pediment gravel above. Williams calculated that present rates of sheet erosion would strip topsoil from the gravel layer within 13 000 years and believes that termites are responsible for its replenishment. Basing his calculations on careful measurements of both size and frequency of termite mounds, Williams concluded that the termites would move 0·47 tons of earth per acre annually and that erosion of abandoned mounds would contribute 3 cm to the topsoil every 1000 years. The concentration of stone in the gravel layer was found to be five times that within the weathered rock beneath, and he calculated that from 12 000–17 700 years would be required to bring about such a stone line 0·3 m thick.

It is clearly difficult to evaluate these analyses, for both pediment formation and termite activity may continue alongside or alternate in a given area. This makes the interpretation of complex stratigraphic relationships in soils hazardous, for although alternating periods of stability and instability are well authenticated for sub-tropical and tropical landsurfaces, it is not yet possible to equate all stone lines with lag gravels on pediments. In fact de Villiers (1965) gives formal recognition to three main formational processes by classifying stone lines:

(1) Those which represent a lag concentrate beneath a creep layer.
(2) Those indicating a surface gravel beneath a colluvial layer.
(3) Those occurring at the sub-surface limit of termite activity.

Summary

It is difficult to summarise accurately the effects of deposition within the tropical zone. In the first place it is a topic which extends into the history of sedimentation that is more a concern of geology, and secondly it involves processes highly sensitive to ecologic changes at any single site. Equally the deposits reflect differences in climate and vegetation cover between sites. The sediments thus tend to reflect the nature of erosional processes discussed in the previous section, and the intensity with which they have operated over varying periods of time and under different climates.

Part Two

The Character and Development of Tropical Terrain

6 The Character of Tropical Terrain

Enquiry into the processes of weathering, denudation and accumulation under tropical conditions has, as its corollary, study of the landforms and deposits which constitute the tropica l landscape. Sometimes there is a close relationship between existing forms and materials and currently operating processes, but in many instances the landforms and deposits have developed under the influence of past climates possessing different character from those of today.

But in either case it is necessary to examine the land form itself, before conclusions can be drawn concerning either the existence or development of characteristic tropical landscapes. In fact some writers have denied the existence of climatically induced morphogenetic regions. This concept of uniformitarianism has been advocated for instance by King (1957) who has argued that under fluviatile conditions hillslope forms vary only in detail, and that landforms such as domed inselbergs result solely from the operation of structural and petrographic factors. Many attempts to define morphogenetic regions have rested almost entirely upon deductive reasoning from a knowledge of climatic parameters, not always, however, known to be causal factors in the fashioning of landscape. Other attempts in this field have relied upon qualitative, and often highly subjective, descriptions of tropical relief. This situation has recently been summarised by Stoddart (1969, p. 174): 'surprisingly little objective morphometric evidence exists on which to base conclusions, and the recognition of distinctive landform assemblages has depended less on total landscape morphometry than on the occurrence of less frequent but more spectacular type landforms, such as inselbergs or pediments.'

In practice, three differing approaches have been made to the problem of analysis and description of tropical terrain. In the first place, as Stoddart points out, individual and distinctive landforms have long received attention. Hypotheses for the development of inselbergs have been advanced, refuted and reiterated over a period of more than seventy years. Secondly, qualitative and often impressionistic descriptions of tropical landscapes are widely encountered in the literature: these commonly have as their aim summary statements of relief type, often designed to distinguish tropical forms from those of other climatic regions (*viz.* Büdel 1957). Finally, a small but growing number of quantitative studies attempt to convey essential, measurable characteristics of form that can be compared from place to place in an objective manner.

Each of these approaches will be examined, but it is not yet possible to assess the results of the more recent quantitative work. There are two basic reasons for this: first, numerical studies are few in number and depend upon mapped data of widely varying quality; secondly, it is important to recognise that the information derived is different in kind from that obtained by other methods. Selected elements of form are measured, and the choice of these is dependent upon the nature of available topographic maps, although some work based upon air photo analysis is emerging. These measures provide new information about drainage texture and relief, but about little else as yet. It is also unfortunate that, because the data on which they have been based varies and cannot be standardised, the comparability of the few studies we do have must remain in doubt.

This reliance upon purely morphological data is severely limiting, but quantitative data concerning surficial materials or hydrology for extensive areas of tropical landscape are almost entirely lacking. It is an essential feature of geomorphological research that, whilst fairly crude measures of relief and drainage composition may be derived from conventional topographic maps, to which some deductions about soil and drainage conditions may be added from aerial photographs, information concerning the material involved in the formative processes must necessarily be sampled in the field. Because of the practical difficulties involved this is confined to more restricted areas. It is therefore seldom possible to relate one set of data with the other.

Many of the general discussions of tropical terrain such as those by Büdel (1957, 1965, 1970), Birot (1968) and Thomas (1965) have been concerned with shield areas such as South America, Africa and India. The prolonged tectonic stability of these areas probably makes them especially suitable for the study of tropical weathering and its effects upon landform, but they cannot be taken to represent the whole of the intertropical zone, much of which has experienced recent crustal upheaval. It is also perhaps unfortunate that the shields of the northern latitudes, which might offer a different kind of comparison, have been the centres of recent glaciation (White 1972); the prolonged effects of a cool or temperate climate upon a stable landmass are therefore little known. Studies of the old Hercynian massifs of Europe confirm the presence of deep chemical weathering of rocks, but also lead to the conclusion that much of this regolith has resulted from the tropical or sub-tropical conditions of the early Tertiary (Jessen 1938; Büdel 1957; Bakker and Levelt 1964; Gellert 1970; Dury 1971).

Studies of individual landforms, however, have been more numerous than morphometric analyses of entire landscapes and much can be learned from these. It is necessary first to qualify certain generalisations concerning the landforms of the tropics. According to Büdel (1957; figure 3, p. 6) the tropical zone is one of plain formation, whilst the cold and temperate areas are zones of valley development resulting from vigorous stream erosion which he identifies as an aftermath of recent glaciation. However, our knowledge of relief development and valley form within the tropics is sufficient to demonstrate a considerable diversity of form. Descriptions by Büdel (1957, 1970) of tropical plains having almost identical longitudinal and transverse valley gradients of from 1·5–3·0 per cent, apply only to very restricted areas of the present-day tropics.

Relief characteristics

In a series of recent studies of the relief of Johore, West Malaysia, Eyles (1969, 1971) and Swan (1967, 1970 a and c, 1972) have built up a clear picture of this area of comparatively stable, and largely granitic terrain. Swan's findings are significant: 'the landscape consists of a bottom storey comprising extensive zones of deposition above which rises a middle storey of dissected lowlands of gentle to moderate steepness with a relative relief within 61 m. A top storey of steep-sided mountains, hills and ridges with occasional plateaux and high plains rises above the lower storeys' (Swan 1970 *c*, p. 101).

This description is substantiated by figures summarised in table 11 below.

TABLE 11

RELIEF CATEGORIES IN JOHORE, WEST MALAYSIA

Relative relief (metres)	*Mean slope (degrees)*	*Percentage area*	*Area of occurrence*
0–50	<1	12·0	coastal sedimentary areas
15–30·5	1·0– 2·0	9·4	alluvial plains in lower river basins
30·5–76	2·0– 4·5	47·0	dissected granite lowlands
76–152·5	4·5– 9·5	9·7	ridges and valleys with occasional monadnocks
152·5–305	9·5–18·5	13·0	dissected hills
305	18·5–34+	8·0	discontinuous mountain areas

(After Swan 1970)

From these figures it can be seen that only 21 per cent of Johore has a relief of less than 30 m and most of this is due to recent, coastal sedimentation. On the other hand 47 per cent of the landmass exhibits mean slope values greater than 2° and relative relief of 30–75 metres.

These observations are corroborated by studies of drainage characteristics from the whole of West Malaysia. Eyles (1971) analysed a stratified 3 per cent sample totalling 410 fourth order drainage basins, amongst which he identified 6 major groups, on the basis of scores allocated in relation to the mean values of selected parameters: hypsometric integral, average slope, basin relief, drainage density and basin area. These groups he described as:

(1) High mountains
(2) Low mountains
(3) Deeply dissected low mountains
(4) Isolated steep high hills
(5) Isolated steep hills
(6) Low convex hills

A fundamental difference was observed between groups 1–3 and groups 4–6, particularly in terms of average slope, basin relief, and drainage density. Between groups 3 and 4 these contrasts can be expressed as:

Average slope: difference 1·32 standard deviation units or 8·3°;
Basin relief: difference 0·75 standard deviation units or 223 m;
Drainage density: difference 1·18 standard deviation units or 3·44 kilometres per square kilometre (5·5 miles per square mile).

Eyles concluded that his figures express a fundamental difference between mountain and lowland basins with an absence of foothills. Eyles (1969) also found that 66 per cent of the basins sampled have a hyposometric integral of less than 34 per cent which Strahler (1952) described as 'the transitory monadnock phase', and which indicated to Eyles (1969, p. 31) 'the high incidence of isolated prominent landforms'. Eyles (1971) also found that whilst basins of groups 4 and 5 were developed mainly over sedimentary rocks, many lowland granite outcrops fell within class 6, suggesting that 'granite weathers to low hills with unusually pronounced summit convexity, which are usually separated by flat floored, swampy stream courses' (Eyles 1971, p. 467). Granites also form many of the isolated and prominent landforms, as well as the mountainous backbone of West Malaysia, so that within a single rock type several distinct types of terrain are found. (Plates 13, 14.)

The abrupt transition from deeply dissected erosional terrain to adjacent alluvial plains is demonstrated even more forcibly in the tectonically active island of Bougainville (New Guinea), where Speight (1967, 1971) found slopes grouped into hillslides at 32° ±10° and alluvial flats with a mean slope of 0° 50′.

Few comparable studies can be cited from other areas. Over the forested basement of southern Nigeria, Jeje (1970) found that dissected lowlands surmounted by isolated inselbergs and quartzite ridges dominated the terrain. These lowlands with less than 60 m of relief occupy more than 60 per cent of the area, and locally 50 per cent of the terrain exhibits relief values of from 16–30 metres.

In a broader view, dissected plains of both lowland and upland character have a very wide distribution over the tropical shields and possess some characteristic valley forms (see p. 145). Sharp breaks in relief and slope values between mountainous terrain and dissected lowlands may be analagous to the widely observed piedmont angle around individual hills. In some cases this break of relief is erosional in character, but elsewhere it marks the juxtaposition of erosional and depositional terrain, particularly in the tectonically active zones.

Over the relatively stable and lowland terrain, *in situ* accumulation of weathering products contributes to a subdued relief and to less distinct and more local distributions of residual and depositional materials. The case represented by Johore illustrates perhaps a transitional situation towards increasing landscape senility, and the unfashionable notion that a cycle of planation is in course of completion may be an unavoidable conclusion. Such instances also recall the comments made by Ruxton and Berry (1961a, p. 28) concerning the occurrence of 'two storey landscapes with upland remnants surrounded by major hillslopes leading down to dissected valley-side strips mantled with laterite profiles'. Some implications of this observation will be explored subsequently, but it may be emphasised here that dissection of the lower storey landscape—corresponding to the middle storey or dissected lowlands of Swan (1970c, 1972)—is in part conditioned by the existence of deep weathering profiles in the humid regions. In more arid areas, shallow weathering and development of a clay plain can be demonstrated (figure 25, p. 105).

The occurrence of marked escarpments or major hillslopes between levels of planation is a recurrent theme in geomorphic literature on the tropical zone, and the nature of the plains above and below such escarpments is no less important. At least in part, these are piedmont slopes extending from the base of major hillslopes and residual hills. In Malaysia, Swan (1970a, p. 38) has concluded that many of these 'though in part colluvial, are essentially residual features, as indicated by the occurrence of corestones, core boulders, and boulders rooted in bedrock within and upon them'.

Drainage characteristics of the tropical zone

Complementary to the study of relief is the analysis of drainage composition, which can convey a clear picture of the texture of terrain and may also express certain hydrological conditions of importance. However, data on drainage texture for the tropical zone are scattered and inconclusive. Comparison of figures from different sources is hazardous because methods for obtaining the drainage net from which measurements are subsequently made differ from author to author. Use of maps of varying scales and degree of detail, or aerial photographs taken at different seasons, produce widely divergent results, especially in the tropics where seasonal stream flow is a marked characteristic and the definition of drainage lines not always without ambiguity. Thus, figures such as those given by Selby (1967a) show that measures of drainage density may vary by a factor of 10^3, and are very difficult to compare (see Doornkamp and King 1971). They also indicate that lithology and relief development, as well as climate and vegetation cover, are major controls. Thus, in semi-arid areas drainage density over poorly consolidated shales may rise into hundreds of kilometres of channel per square kilometre of terrain, whilst in humid temperate areas with a forest or woodland vegetation cover, figures tend to vary from 1·8–5·6 km/km^2 (Selby 1967a).

Working in the western United States, Melton (1957) found a wide range of densities from 3–187 km/km^2, and he considered that these figures were influenced by climate, infiltration capacity and vegetal cover. Of these factors only infiltration is directly affected by lithology, and all are closely related to climate. Chorley (1957) and Chorley and Morgan (1962) compared three lithologically similar areas from Britain and the United States and confirmed a close relationship between drainage density and rainfall. The areas examined were uplands and their figures ranged from 2 km/km^2 on Exmoor to over 7 km/km^2 in the Unaka Mountains, Tennessee.

It is perhaps significant that figures from areas of low relief in the humid tropics generally exhibit a range below those just quoted. Doornkamp (Doornkamp and King 1971) has analysed more than one hundred, third-order basins from Uganda, spanning a wide variety of relief types. Densities ranged from 0·62–6·25 km/km^2, but 54·7 per cent of the basins exhibited densities of from 1·25–2·5 km/km^2, and those which were higher fell within a group of hilly catchments, having basin relief figures of 256–345 m, and local valley relief as high as 213 metres. Low drainage densities within the range 0·5–1·5 km/km^2 have been found widely over the Nigerian basement, from the rain forest boundary and the wetter savannas (Wigwe 1966; Jeje 1970) to as far north as Zaria which experiences a highly seasonal climate (Thorp 1970).

Such low densities probably reflect the widespread occurrence of deep regoliths with infiltration capacities high enough to absorb a large proportion of the precipitation (Nye 1954). Such regoliths have been found to extend well into the savannas (Thomas 1966a) and they possibly reduce any tendency for drainage densities to increase as precipitation intensity and length of dry season increase, and vegetation cover becomes reduced. The high rates of denudation predicted for this zone by Fournier (1962, figure 26, p. 115) might be taken to imply high drainage densities, but these do not appear to occur on terrains of low relief.

A rapid rise in drainage density occurs, with increasing relief in the tropics. This is clear from the work of Doornkamp and King (1971) who found a maximum of 6·25 km/km² in Uganda, and particularly from the studies made by Eyles (1971) of drainage basins in West Malaysia, where densities rise to 12·5 km/km² in mountainous catchments. However, the drainage density over the whole peninsular appears to be much higher than that found on the old, basement landsurfaces of Africa. These findings may be related to higher relief figures and thinner regoliths in Malaysia, as well as to a generally higher annual rainfall. It is interesting that Peltier (1962), in a pilot study undertaken from the analysis of large numbers of topographic maps, found that there was a much steeper rise in drainage density with increasing mean slope in tropical areas than in any other (non-glaciated) climatic zone. Since increasing mean slope will usually be accompanied by increases in absolute altitude, local relief and valley side slopes, it is likely that total precipitation amount and regolith type and depth will be major factors accounting for these figures. In semi-arid or temperate areas, such variations may be less.

However, it is dangerous to attempt firm generalisations from these data and it is also necessary to recognise that measurements of stream lengths and basin areas cannot alone summarise the hydrological character of a region. We require in addition the dimensions of stream channels, and their bankful discharges, combined with hydrographs representative of the areas studied. Without such information it is doubtful whether we can progress very far with analysis of drainage characteristics in the tropics or elsewhere. Added to this is the widespread but uneven influence of man on the vegetation cover, which has continued for many centuries. Drainage density is likely to be increased by depletion of vegetation cover, whilst older slope forms and deposits in the landscape survive from remoter periods. Ruxton (1968) has emphasised that stream channels adjust to changes of environment quite rapidly, but other aspects of the landform may be inherited from a more distant past and we cannot hope to interpret the present-day landscape solely in terms of contemporary dynamics. These different components of the landscape are difficult to analyse and possibly overshadow the immediate question of providing comparable data on drainage texture from more numerous locations.

LANDFORMS OF THE TROPICAL ZONE

The concept of the 'landform' is not without difficulty, for although firmly embedded in the literature of geomorphology, the term is used indiscriminately for component features of the land surface which vary in both morphological and genetic complexity. Savigear (1965) recognised the need for definition and suggested that, 'a landform is a feature of the earth's surface with distinctive form characters which can be attributed to the dominance of particular processes or particular structures in the course of its development and to which the feature can be clearly related' (p. 514). Even this definition leaves the question of scale as a matter of ambiguity, although size is to some extent dependent upon the controlling structures or processes and their areas of occurrence or operation.

The most obvious and nearly ubiquitous landform is the river valley, and attempts have been made to classify valley forms on the basis of transverse profiles. Nevertheless, it must be admitted that valley profiles can only properly be understood within the drainage basins of which they are part. Less ambiguous are the discrete forms of the land surface: individual hills and units of slope. These fall naturally into three groups: those developed over the weathered mantle, those formed from solid rock, and slopes developed over transported materials. But of course many hills that must be recognised as 'landforms' may be comprised of many individual slope 'facets' that may have formed over any of these materials. Even the rocky inselbergs of the tropical shields possess a considerable variety of surface materials (Thomas 1967), and the controversial 'pediment' landform may also be underlain by regolith, rock or alluvium.

Attention has nearly always been focussed in geomorphology upon residual hills, and valley forms as the most clearly defined 'objects' in the physical landscape. Between these two, fall the pediment and related piedmont features which extend from the hillfoot towards the adjacent or distant valley, and it may be for this reason that the pediment has created so much difficulty in the study of landforms. Escarpments also occupy a central position in geomorphic study, for although they have much in common with the marginal slopes of residual hills, they are frequently more complex, and are often assumed (not always correctly) to link upper and lower landsurfaces of different age.

It is also manifestly true that individual landforms which may be rather arbitrarily defined must finally be understood in relation to present or past systems of interconnected forms, linked by processes of weathering, water flow and mass movement. It is the purpose of the following sections to examine in some detail the character of slope and valley forms, the rocky hills of the inselberg landscape, and the nature of pediments and related forms within the sub-humid and humid tropics.

Slope and valley forms

Attempts have been made to characterise tropical landsurfaces according to slope and valley forms. Louis (1964) for instance has proposed a classification of valley forms based on observations from the old planation surfaces of Tanzania (figure 27). Over the largely intact 'peneplains', 'pure saucer-shaped valleys' (*Reine Flächmüldentaler*) predominate. These have broad flat, alluviated floors and gently concave slopes underlain by deep red soils. These valley-side slopes of generally less than 3° inclination were called ramp-slopes (*Rampenhang*) and were of considerable length. These were found almost exclusively within the savannas where rainfall is highly seasonal in distribution and totals from 500–1000 millimetres.

With higher rainfalls, groove-shaped valleys (*Kehltaler*) possessing alluviated concave floors confined within steep valley-side slopes were found, whilst another departure from the saucer-shaped form was the incised valley form with convex slopes and commonly with a flat floor (*Sohlenkerbtaler*). Louis (1964) associated these valley forms with climatically controlled planation, and found that their occurrence was largely independent of altitude. Few tests of this largely qualitative scheme have been attempted. Wigwe (1966) examined more than 100 valley profiles in an area of Nigeria some 60 kilometres north of the forest boundary and close to the main

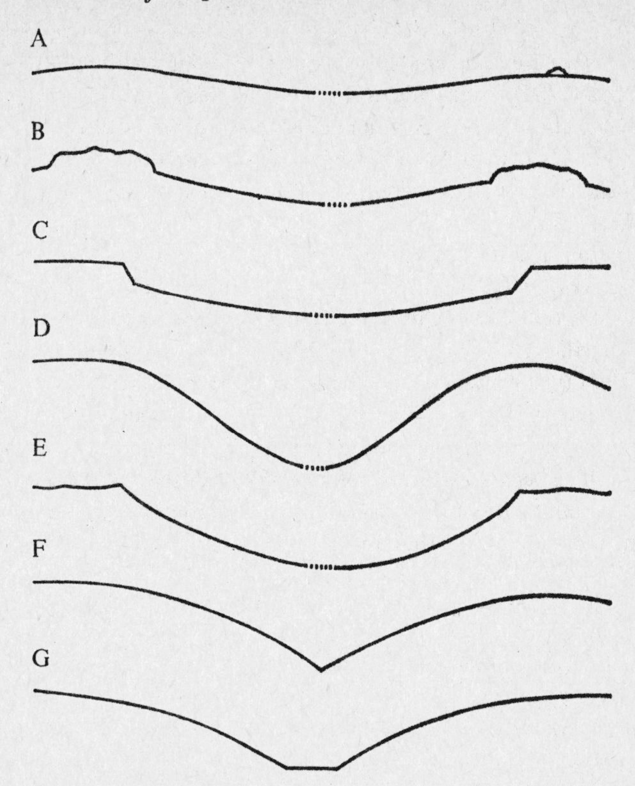

Figure 27 Schematic diagrams to show typical valley forms in Tanzania (after Louis 1964).

A. Pure saucer shaped valley (Reine Flächmüldental); B. Saucer shaped valley with ramp slope in crystalline rocks (Flächmüldental mit Rahmenhöhen); C. Saucer shaped valley with ramp slope in sandstones; D. Groove shaped valley in crystalline rocks (Kehltal); E. Groove shaped valley in sandstones; F. V-shaped valley (Kerbtal); G. Steep-sided valley with flat floor (Sohlenkerbtal).

drainage parting between southward flowing. Atlantic drainage, and streams flowing north into the Niger. This is an area of laterite uplands over basement rocks, dissected to varying degrees and surrounded by landsurfaces from which varying amounts of weathered material have been removed by erosion. In the laterite uplands he described 'bowl-shaped valleys' possessing extensive ramp slopes leading upwards to the debris slopes and breakaways of the laterite hills. Elsewhere 'saucer-shaped' valleys similar to those described by Louis (1964) cover wide areas. These display greater convexity of summit areas, but possess long concave ramp slopes of less than 3°, and wide marshy floors across which flow small streams in multiple channels. Wigwe also recognised some areas of V-shaped valleys like *Kerbtaler*.

In a recent study of valley forms in the Matto Grosso area of Brazil, Young (1970) has noted the predominance of convexity amongst profiles from within the rain forest, and has called attention to similarities with the *Sohlenkerbtaler* of Louis. He found the *Flächmüldental* type only in valley heads where gallery forest was present along the stream lines. On the basis of more than 80 measured profiles, Young (1970) proposed five major classes of slope form (figure 28). There is a striking absence of level or indeterminate slopes, even on gently sloping interfluves, and a predominance of convexity in an area straddling the forest savanna boundary, underlain by

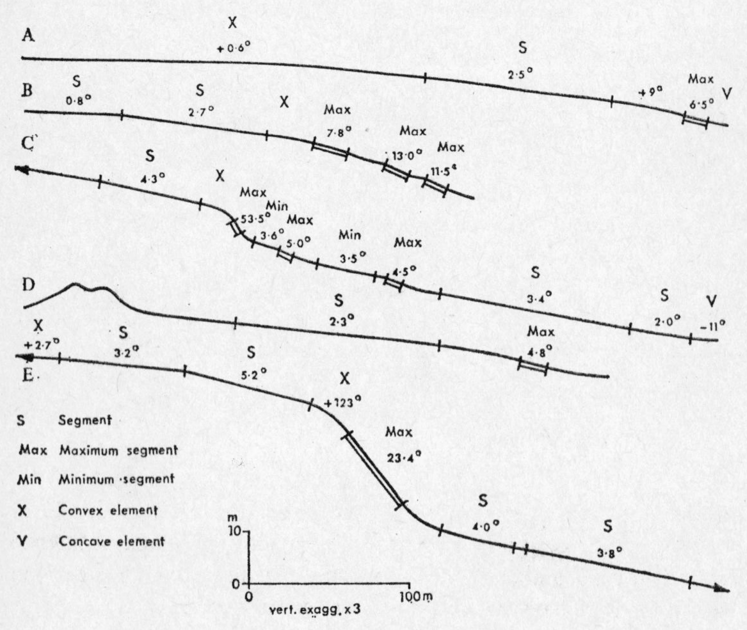

Figure 28 Slope forms in the Xarantina–Cachimbo area of Brazil (after Young 1970).

sub-horizontal sandstones and shales and dissected to depths of 20–40 metres. Young considered slopes of categories A to C (figure 28) to be a developmental sequence. Incision of the stream into the smooth-crested, unincised valleys leads to the development first of strongly convex, incised forms, and then to the development of benched valley-side slopes showing upper convexity, below which the slope steepens to over 45°, before levelling out towards the valley floor. Other slopes were surmounted by small laterite hills below which extended linear slopes of 2–3° inclination and up to 400 m in length (figure 28D). Elsewhere a strong laterite scarp was developed, inducing a slope of 23–25°, below which was an abrupt break towards the stream (figure 28E).

Valley forms in laterite terrain from north-eastern Brazil were also discussed by Vann (1963) who described wide marshy floors, bordered by steep, spring sapped slopes. Eyles (1968) has also pointed out similar valley forms under forest in West Malaysia. Here, massive laterite does not occur widely, but the broad marshy floor, often with little sign of a permanent channel, again abuts against a steep slope rising towards the broad interfluve surface.

Such river valleys with flat marshy floors but lacking in well developed stream channels appear very commonly within the seasonally humid tropical plainlands. In Central Africa they are commonly called 'Dambos', and they are generally developed across deeply weathered and often lateritised terrain (Webster 1965). Although such valleys may be confined within marked valley-side slopes, they often appear to function as little more than concentrated zones of sheet wash. Young (1969) has described very broad and open valleys of this kind, from plains lying between 1000 and 1400 m in Malawi, where the valleys were up to 1600 m wide with maximum slopes varying from 2–8°. The valley floors of the smaller tributaries seldom contain a defined stream channel, but exhibit a gently concave cross section and a cover of swamp grassland. These conditions are associated with a four month wet season (mean annual rainfall about 1000 mm) and a period of one or more months during which no flow is evident within the smaller stream valleys. Such conditions of negligible dry season flow were described also by Ledger (1969) from West Africa and it appears that many such valleys are deeply weathered below the level of the valley floor. Where steep flanking slopes exist these seem to have developed by spring sapping of laterite deposits as described by Vann (1963). The very widespread influence of laterite upon valley form is nearly always to produce scarp and bench features, resulting from the rapid induration of the deposit wherever it is exposed on the valley flank as a result of stream incision. Frequently, in West Africa valley forms exhibit an upper convexity bordered by a steep laterite breakaway, below which a 'pediment' slope extends towards a bench or ledge of secondary laterite, beyond which the slope steepens again towards the stream channel (De Swardt 1953, 1964; and figure 14, p. 64).

There are, in fact, many problems both practical and theoretical in the description of valley forms, especially in well planated, tropical areas. In the first place many of the valleys are very extensive features, requiring very long traverses in the field, and methods for comparing several such transverse profiles are not clearly established. Objective methods for the description of slope units within the valleys have been advocated (Young 1970; Savigear 1967), but have not yet been widely applied. As with measures of drainage texture, the morphological facts are of limited use without a knowledge of stream hydrology, and particularly in the case of slopes, a knowledge of the nature of the underlying soil materials. In many cases, the valley side is identical with the concept of the pediment, and further comment on these features is given below (p. 215).

What is clear, though we have few figures to substantiate it, is that valley forms vary fundamentally with lithology. Forms developed on lateritised regoliths, and over sub-horizontal sedimentary formations are quite different from those occurring over stripped surfaces of gneiss or granite. The former are quite often deeply incised, and although they may have broad, flat floors, commonly infilled with alluvium, their maximum slopes may be steep, exceeding 25° quite commonly. On the other hand wide, open valleys tend to develop over the acid gneisses. In fact in some areas quite vigorous rivers appear to make little headway in deepening their valleys. Instances of this come mainly from the savannas, where rivers flowing over solid rock are visibly both corroding and abraiding the channel floor. Yet the flanks of the valleys rise gently over shallow soil profiles, suggesting that slope erosion and channel erosion are closely balanced. But these are shield situations, where gentle tilting of possibly ancient planation surfaces may have led to forms quite untypical of other areas. For instance, in Malaysia the granites

Plate 1 Laterite profile over basic metamorphic rock, near Kano, Nigeria.
The duricrust capping is 3–5 m thick, and the complete profile is *c.* 40 m; cores of unweathered rock are seen in the foreground.

Plate 2 Laterite mesa on the Jos Plateau, Nigeria.
Granite cores seen in the foreground.

Plate 3 Marginal disintegration of duricrust cap in the Nigerian savanna zone.

Plate 4 Deep weathering in Precambrian gneiss, near Sekondi, Ghana.
The basal surface of weathering has been artificially exposed during quarry operations; weathering depth is 20–30 m.

Plate 5 Weathering profile with corestones in granodiorite, Singapore Island.

Plate 6 Deeply weathered interfluve in Precambrian gneiss in the forest zone of southern Nigeria.

Plate 7 Landslide scar over basic intrusive rocks of the Freetown Peninsula, Sierra Leone.

Stripping of the basal surface is evident, with accumulation of slide material in the foreground. Water streaming from remaining regolith on upper slopes suggests that lubrication of the basal surface may have induced the slide. Annual rainfall at the site is in the region of 3000 mm.

Plate 8 Colluvial laterite debris overlying a weathered landsurface in central Sierra Leone. Annual rainfall exceeds 2500 mm.

Plate 9 Slump scars and earthflows on slopes cleared of rain forest, over weathered schists in the Owen Stanley Mountains, Papua.

Similar if more subdued forms can be recognised beneath the forest canopy. The flat alluviated surface in the foreground has probably resulted from back tilting of the drainage system during uplift.

Plate 10 Corestones accumulated in the floor of a small stream valley, in granite on the Jos Plateau, Nigeria.

The stream is concealed at a depth of 5–10 m below the surface at this point.

Plate 11 'Stone line' beneath a shallow soil cover, on a weathered landsurface in the Nigerian savanna zone.

Plate 12 Deeply buried stone line in the forest zone near Kumasi in Ghana.

Plate 13 Rain forest developed over a thin weathering profile on granite in Johore, West Malaysia.

The surface is strewn with granite cores, but seldom completely stripped, even where slopes reach 20–30°.

Plate 14 Valley form in dissected granite country, Johore, West Malaysia.

Although deeply incised, with long straight slopes, the valley floor is flat and without a marked stream channel. It is underlain by alluvium (see also Swan, 1972).

Plate 15 Tor groups in granite near Zagun on the edge of the Jos Plateau (see also figure 31).

The main tor group shows a domical form, suggesting initial formation of a large domed mass, later sub-divided by groundwater weathering into the existing form. On the left a similar group has apparently collapsed outwards as a result of further weathering (see figure 34).

Plate 16 Deep weathering adjacent to degraded tor form in granodiorite on the Snowy Mountains Highway, New South Wales.

It seems likely that the weathering extends beneath the tor, which is really no more than a group of corestones (see figure 34).

Plate 17 Domed inselberg in granite near Panyam on the Jos Plateau, Nigeria. This small hill is nearly symmetrical, and shows a single, separated, convex rock sheet.

Plate 18 Asymmetrical domed inselberg in Precambrian gneiss, near Oyo, Nigeria. A seasonal stream flows along the foot of the hill on one side; the valley slope continues to rise behind the hill (see figure 33).

Plate 19 Domed inselberg in the rain forest of central Sierra Leone.
Some writers interpret such vegetation patterns as an indication of colonisation and progressive concealment of rock domes in the forest zone.

Plate 20 Domed inselberg in the savanna zone, near Funtua in Nigeria.

Plate 21 Stripping of cores and a domed rock surface in a valley head on the Jos Plateau, Nigeria.

Plate 22 Coarse waste material resting against the side of a domed inselberg near Kusheriki in the savanna zone of Nigeria.

Runoff from upper slopes rapidly percolates into this mantle but may also lead to further stripping. Weathering beneath the debris may bring about basal sapping of the dome surface (see figures 34, 39).

Plate 23 Stream channel developed across a convex surface of granite near Sha on the Jos Plateau, Nigeria.

The river appears merely to flow over the contour of the dome, becoming incised only along fracture planes, and around separated rock sheets.

Plate 24 Granite escarpment around the Kagoro Hills, Nigeria.

The upper parts of the slope show evidence of both sub-surface weathering and tensional collapse. A talus of corestones and joint blocks is concealed by forest at the base of the hillslope.

Plate 25 Stripped footslopes below the Jos Plateau escarpment near Toro, Nigeria.

Powerful stripping from sheet floods is actively removing a shallow, lateritised regolith over granite.

Plate 26 Backwearing along sub-horizontal exfoliation sheets forming granite footslopes to a granite escarpment near Toro, Nigeria.

are deeply dissected by valleys and the stream channels carry quite large boulders: a point made concerning some of the rivers in New Guinea by Speight (1968), and borne out by the relief characteristics for Johore previously quoted.

Residual hills of tropical landscapes

Prominent among discussions of tropical landforms, are accounts of inselberg landscapes (*inselberglandschaften*) within which prominent, and often isolated hills rise steeply from plains, like islands from the sea. Although Young (1972) has recently warned against the misapprehension that the entire inter-tropical zone is dominated by such landscapes, they remain of sufficient importance to merit extended study in the present context.

The term 'inselberg' appears to have been proposed by Bornhardt in 1900, to describe the abrupt and rocky hills that commonly interrupt tropical plains. But use of the term was soon widened to include a variety of isolated hill forms. W. Penck (1924) described the residual hills rising above the plains of Saxony and Franconia as *inselberge,* and much later, King (1953, 1957, 1962) extended the term to describe any steep-sided residual hill that had arisen from the operation of the twin processes of scarp retreat and pediplanation. Thus, both descriptive and genetic connotations attach to the term and its usage has become ambiguous. There is, however, little merit in attempting to replace the term inselberg, and because it is descriptive of a wide range of isolated hill forms it will be applied here generically and descriptively but without genetic connotations.

Inselbergs (or *inselberge*) may therefore be taken to include hills of sedimentary rock, residuals of saprolite capped by indurated laterite, and hills of crystalline rock which vary from tors or 'boulder inselbergs' (Birot 1968) and angular castellated forms or 'kopjes' (or 'koppies'), to the massive rock domes originally described by Bornhardt himself (see plates 15, 17–20).

In order to distinguish this last type of hill clearly from other groups of forms, Willis (1936, p. 117) proposed the adoption of Bornhardt's own name to describe hills with 'bare surfaces, dome-like summits, precipitous sides becoming steeper towards the base, an absence of talus, alluvial cones or soil (and) a close adjustment of form to internal structure'. Thus, such forms have often been described in English as 'bornhardts', particularly by King (1948), and also by myself (Thomas 1965a, 1967) among others. However, the widespread use of the qualified term 'domed inselberg' (*Domförmiger Inselberg* in German, and *dôme crystalline* or *dôme granitque* in French), is perhaps to be preferred for its descriptive accuracy.

Some problems of terminology still remain, however, because multi-convex rock forms in crystalline rocks vary from low rock pavements which barely break the surface of the plains, to domes of varying size and morphological complexity. There is also a need to distinguish the spheroidal forms of boulder inselbergs, or tors, from domical forms and this poses some difficulties. Furthermore, not all residual granite hills fall into these categories; some remain mantled by a thin regolith interrupted here and there by tors and half domes. According to the analyses of Malaysian relief just discussed, these are also residual hills.

The distinction between tors and domes is usually made in terms of the marked spheroidal form and generally smaller size of the former. But in reality spheroidal and domical forms are not mutually exclusive. Large tors may not have their undersides exposed, while major domes may exhibit incurved basal slopes. Cunningham (1971) has suggested that tors are characterised by horizontal jointing, while domes commonly exhibit curvilinear sheeting. But this is not always so: many tors appear to arise from the separation and weathering of curved sheets of rock. It is possible only to state that spheroidal tors are generally within 10^0 or 10^1 m in diameter, while domes vary upwards from 10) to perhaps 10^2 or 10^3 in diameter (Thomas 1974). Possible distinctions on genetic grounds between these various sizes of form are discussed here.

It is often assumed that from the base of all residual inselbergs will extend gently inclined piedmont surfaces ranging from rock pediments, through weathered slopes to depositional surfaces (*glacis*), comprised of colluvial and alluvial formations. Characteristically this is so, and such broad and gently sloping surfaces pose many problems of description and definition for the geomorphologist. However, many hills that would be described as inselbergs occur in closely spaced groups or emerge from steep hillsides, and generalisation is again difficult.

The low rock pavements which punctuate the surface of the plains have been variously described as 'ruwares' (King 1948; Thomas 1965a) and as isolated wash pediments (*Isiolertes Spülpedimente:*

Büdel 1957), but 'rock pavement' appears to avoid the linguistic and genetic problems which these alternatives pose.

The domes and tors which do not occur strictly as inselbergs, commonly form parts of escarpments which form major features within many of the landscapes of the tropical shields. Büdel (1957), Louis (1964) and Pugh (1966) among many other authors have stressed that tropical plains are commonly bordered or terminated by escarpments. It is also part of the argument of those who contend that tropical streams are unable to achieve significant vertical erosion that such escarpments reflect this situation. Many escarpments are in fact products of differential erosion, and most are complex, containing many individual landforms.

It can perhaps be seen from this short section that outside of the tectonically active zones certain recurrent features of tropical terrains may be described in terms of characteristic valley forms, and residual hills. The 'landscapes' which these comprise are discussed in a subsequent section and the evolution of specific forms and their importance to an understanding of tropical denudation are also considered later at some length.

Residual hills of karst landscapes

Before exploring these themes further it is important to recognise that not all residual hills in the tropics occur as tabular forms in weathered and sedimentary materials, or as domed hills of crystalline rock. In focussing attention upon the more widespread landscapes of the silicate rocks and detrital sedimentaries, there is a danger of neglecting the individuality of terrains developed over other rock types. Distinctive forms are particularly associated with limestone or karstic terrain. Thus, although this study does not seek to explore the origins and development of tropical karst, a comment on karstic residuals which in some respects resemble other inselbergs is given here.

Residual *Karstinselberge* are common in humid tropical limestone regions, and are characterised by steep, sometimes vertical, sides which may be case-hardened by redeposited, crystalline calcium carbonate. Such hills may represent limestone residuals standing above plains developed across an underlying, relatively impermeable rock or they may be separated by alluviated depressions underlain by less resistant limestones. Such hills are given a variety of local names, though the smaller ones are often termed *hums* and very large residuals are called *mogotes* (Sweeting 1958, 1968). The superficial resemblance of these hills to the domed inselbergs of granitic rocks is of interest, because it is generally agreed that solutional weathering is the dominant process responsible for their formation, and they are largely confined to the more humid regions of the tropics and sub-tropics.

The solutional weathering of limestones is a complex subject which lies outside the scope of this study, but it is important to recognise that the amount of absorbed CO_2 in rainwater is less in warm climates than in colder areas. For this reason the solutional modification of bare rock surfaces is thought to be less in the tropics than in higher latitudes. On the other hand, the amount of absorbed CO_2 in soil beneath a forest cover may rise to 15 times that of the external atmosphere (Sweeting 1968), and these high levels promote rapid solution of the limestone.

Solutional weathering of the limestone penetrates in a dominantly vertical direction, and is favoured by the presence of thick, well jointed limestone formations and by a high available relief. The rate of solution is closely related to rainfall amount, wherever a soil and vegetation cover exists. It appears that tropical karsts develop by stages during which solution hollows or *dolines* may develop under wide watertable fluctuations into steep walled 'cockpits', the floors of which become alluviated. Conical hills commonly rise above these depressions (Sweeting 1958) to form 'cone and cockpit karst' (*Kegelkarst*). Where this development is accentuated, by climatic, lithologic or physiographic factors a 'tower karst' (*Turm Karst*) dominated by vertical walled mogotes rising abruptly from an alluviated plain may be formed. Cone karsts are common in Jamaica (Sweeting 1958) and tower karsts have been described particularly from South China and North Vietnam.

Once the residual hills have become denuded of vegetation cover, and intervening depressions covered with alluvium, the contrast in weathering rates becomes such that, given favourable conditions of relief and climate together with a thick limestone formation, great vertical development within the tropical karst landscape becomes possible.

Some problems of describing karst landscapes have recently been discussed by Williams (1971), who, by means of a careful morphometric analysis, was able to demonstrate a clear organisation within two karst areas in New Guinea. Ridges in one area near Mount Kaijende rise to saw-edged arêtes more than 100 m high at an altitude of 2900 metres. At this height temperatures probably range from 10–13°C with little variation during the year. This is sufficient to nourish a dense moss forest, under an annual rainfall which probably approaches 4000 mm, and yields drainage waters of high acidity (pH 3·9). Jennings and Bik (1962) suggest that such bioclimatic conditions may come close to a world optimum for limestone solution and karst development.

It is clear from these comments that karst development in the tropics offers some intriguing parallels with the inselberg landscapes discussed here in greater detail. But fundamental differences in the process of alteration, and conditions for optimum weathering occur which warn against attempting too close a correspondence between two distinct morphogenetic systems. Two common features of the weathering environments which favour the rapid alteration of both groups of rocks are the maintenance of a soil cover or waste mantle, and the growth of a luxuriant vegetation cover. It appears that in both cases the emergence of bare rock forms leads to a sharp reduction in the rate of weathering with important implications for the study of landform development.

TYPES OF TROPICAL TERRAIN

It is not possible to present a complete and objective typology of land surfaces within the tropics. It has been shown that in terms of relief and drainage characteristics, we have too few measurements to provide a basis for generalisation. Furthermore, many measurements which we do have suggest that few unique or unusual characteristics are evident. This is partly due to the use of conventional techniques that do not necessarily emphasise the most important features of tropical terrain. Many of these concern the nature of surface materials, and in the study of individual landforms their importance will be stressed. Identification of distinctive terrain types should therefore rest less on simple morphometry than on the pattern of occurrence of particular surface materials and landforms developed on them. Such an approach will be followed here, but it should be recognised that it is not considered as an alternative to quantification of terrain attributes, for most of the features mentioned can quite easily be subjected to measurement. Unfortunately few have been measured over extensive areas or for statistically useful samples. As so often in geomorphology, we have tended to leap into the discussion of complex and intriguing problems of origin and evolution of forms, before having any clear idea of what the landsurface is like over most of the earth.

The schemes discussed here are therefore largely qualitative, but they contain some measured or measurable characteristics, and should be capable of being tested and improved. Few published accounts can be compared directly for reasons implicit in these introductory remarks. Nevertheless, recurrent themes occur which permit some useful discussion.

Büdel (1965) recognised three basic relief types in Madras: (see figure 29).

(1) *Tropical plain relief* (Rumpflächen) is exemplified by the Tamilnad Plain which rises from sea level to about 100 m inland, and possesses all the features of extreme tropical planation described elsewhere by Büdel. Slopes developed over deeply weathered rocks attain very low values (1·2–1·7 per cent), but are surmounted by isolated, high inselbergs (*Auslieger Inselberge*) bordered by narrow radial pediments with slopes of 3·5–4·0 per cent. Above a dissected escarpment zone, a similar plain, the Bangalore Plain, is preserved at an altitude of 750–900 metres.

(2) *Tropical mountain relief* forms the inner border of the Tamilnad Plain and is characterised by frequent inselbergs or groups of hills leading towards an escarpment dissected by deep, narrow valleys.

(3) *Tropical ridge relief* appears above the escarpment, and merges into the Bangalore Plain. It consists of narrow groove-like valleys etched into the upland surface by selective weathering and concentrated slope wash.

A similar scheme was suggested by Harpum (1963) for a part of Tanzania, where over an area of granite he recognised *Plateau scenery* comprising extensive surfaces of planation, dominated mainly by regolith covered slopes, and carrying few residual hills. Rock outcrops are confined mainly to stream channels. This is

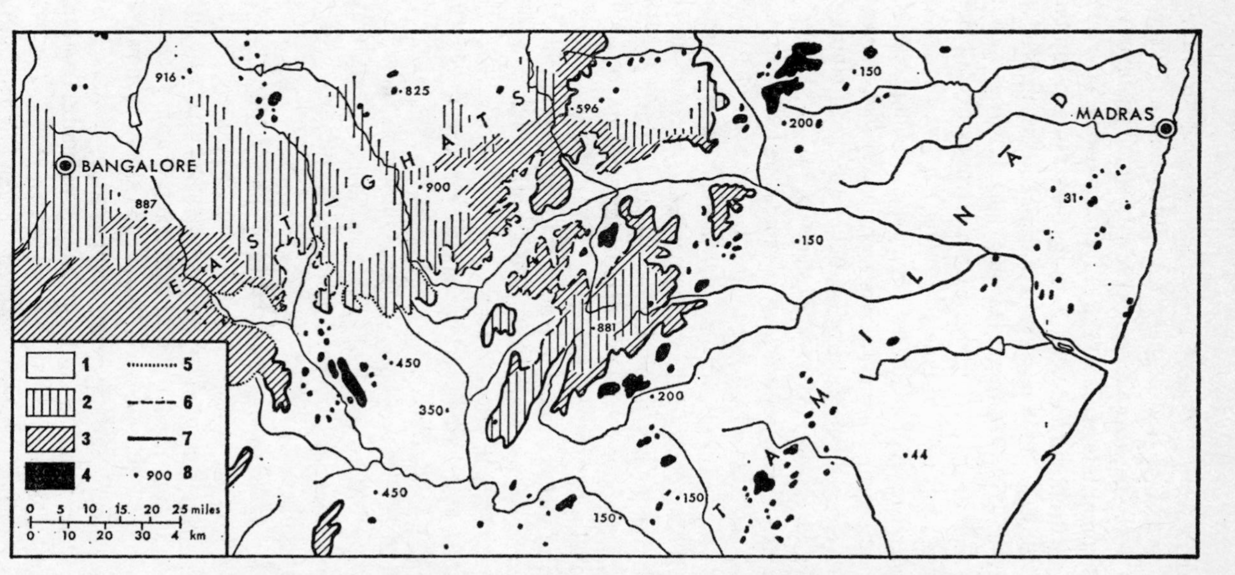

Figure 29 Tropical Relief Types in a part of south India (after Büdel 1965).

1. Tropical plain relief (Rumpflächen); 2. Tropical ridge relief (Ruckenrelief); 3. Tropical mountain relief (Gebirges relief); 4. Inselbergs; 5, 6, 7. Escarpments (Rumfstüfen): respectively poorly defined, well defined, and abrupt; 8. Heights: in metres a.s.l.

dissected into *Hill scenery* displaying a complex of structurally compartmented hills, many of them bare rock domes. This was distinguished from *Inselberg and Tor scenery* in which pediments of from 2–6° surround isolated hills, and the intervening plains are extensively and deeply weathered. In some places dissection was carried below the level of the basal surface of weathering to produce incised, rocky valleys. To these Harpum added *Scarp scenery,* and *Buried scenery* which had resulted from extensive aggradation.

Hurault (1967) has also considered granite relief in the tropics and described landscapes of weathered, cupola-shaped compartments. These he characterised as *Alveolate relief* which persists commonly below escarpments in the humid tropics. Other aspects of Hurault's analysis (1967) are considered later (chapter 9, p. 248).

The occurrence of dissected lowland terrain in the humid tropics, especially over granites, is emphasised also by the work of Ruxton and Berry (1961a) and Swan (1970a), and in many cases piedmont zones have this character. Whether this may be attributed to deeper weathering in the humid tropical zone; to the regular jointing patterns of the granite; to a particular degree or kind of uplift, or perhaps to all three is not clear.

A sequence of surfaces characteristic of many tropical shields has been described by Doornkamp (1968, 1970) from Uganda, where he recognised five principal terrain types:

(1) *Residual hills* forming groups standing above the highest extensive landsurface.
(2) *The upland landscape* carrying many duricrusted hills.
(3) *Inselberg areas* developed over granitic rocks, where the upland landscape has been destroyed by the dissection of deep weathering profiles.
(4) *The lowland landscape* covered by a thinner and less well developed laterite.
(5) *Infill or aggradation landscapes* which have arisen in large measure as a result of drainage reversal and deposition within the Lake Victoria Basin. In Malaysia however, aggradational surfaces were recognised by Swan (1970b and c) and formed a girdle or lowland, around the residual terrain of Johore.

In a similar way, Mabbutt (1962) grouped the Land Systems in the Alice Springs area into four main groups:

(1) *The erosional weathered land surface* comprising the remnants of an ancient cycle of erosion, represented by old weathering profiles and residual groups of hills controlled by lithology.
(2) *The partially dissected erosional weathered landsurface*, where erosion has partially destroyed the older forms by stripping or deeper dissection.
(3) *Erosional surfaces formed below the weathered landsurface*, where only younger forms survive.
(4) *Depositional surfaces*

Since these were recognised from a wide area of diverse lithology, Mabbutt was able to comment upon the influence of rock type on the terrain, and distinguished sub-divisions according to the occurrence of mountains, hills and plains on granite, gneiss or schist; ranges of sandstone or quartzite, and undulating terrain on weathered sedimentary rocks (see also Wright 1963).

A more general statement of terrain condition, based on the intensity of processes operating over the African landmass was given by D'Hoore (1964) who distinguished:

(1) *Surfaces* too level to permit important movements of surface materials.
(2) *Zones of transference* within which soil material was in course of movement towards depositional environments.
(3) *Zones of accumulation*, formed especially within the interior basins of the continent.
(4) The *rejuvenated zones* of Africa, found particularly around the uplifted rim of the eastern and southern part of the continent.

This scheme embodies the concepts of landsurface stability and instability previously discussed (p. 124). Clayton, working in Nigeria (1958), also referred to *stable surfaces* over which lateritic weathering profiles were found, and which are separated by escarpments or narrow zones of dissection and transfer of earth materials.

The survival of stable landsurfaces for long periods of time in the tropics, is clearly a situation leading to the production of deep regoliths and the formation of distinctly zoned laterite profiles. The extreme weathering of soil parent materials and the development of distinctive landforms within the regolith material, both contribute to many of the characteristics of tropical plains that have been emphasised in this study, and which appear to mark them off from temperate or cold climate forms.

In some cases a classification of surfaces by known geological age may be a guide to the development of these features, but in fact the recognition of such planation surfaces often depends upon the occurrence of supposedly diagnostic features such as laterite, and the argument becomes circular. Altitudinal differences between plains are in fact a poor guide to the *actual* age of the landsurfaces and their weathered mantles. Thus, in northern Nigeria similar lateritic weathering profiles may be found both above and below granite scarps, and the inference must clearly be that, whatever the age of the initial formation of the two plains, their present form and materials have resulted from a common experience of prolonged weathering and a low rate of surface denudation. Obviously if, as is commonly thought, the middle Tertiary period in Africa was one of prolonged, humid tropical weathering, then any surfaces developed prior to that period would be similarly affected, unless already dissected or uplifted to very high altitudes. It is this situation that has led to much of the confusion over the dating of landsurfaces in tropical Africa (Dixey 1955).

An approach to landsurface classification which is based upon the occurrence and pattern of certain landforms and landforming materials should offer a rational approach to these problems. To some extent this involves problems of definition, but it is broadly possible to distinguish slope *facets* of particular materials, range of slope and drainage conditions (Beckett and Webster 1965), and we have a vast literature in which the assumption is made that 'landforms' as recognisable entities occur repeatedly over the earth's surface. Such an approach builds into an hierarchical system of ordering in which facets (which can be further subdivided) are combined into landforms, and landforms are seen to be repeated in characteristic patterns or Land Systems (Christian 1957, 1968; Brink *et al.* 1966). Such systems have usually been defined *ad hoc* and in local terms, but in principle broad landform types, from which there will be many and important local variants, should be capable of definition for at least related groups of landforms, such as those of tropical shields, sub-horizontal sedimentary rocks and other structural forms.

This approach has the advantage of using the 'building block' principle in which fundamental units of terrain are first defined and then recognised in characteristic combinations throughout areas of similar relief type. Such a method is capable of embracing a wide range of terrain properties and does not emphasise particular elements of form, such as drainage lines, relief or slope. In particular it lends itself to the recognition of particular classes of slope materials. At the same time quantification can be incorporated into the definitions by representing them in terms of the frequency, position and area of occurrence of the components or slope facets within the system. Such statements can also be combined, without procedural difficulty, with morphometric summaries of terrain relief and drainage texture.

A table is given of common landform facets within tropical landscapes (table 12). It is not suggested that this is exhaustive or sufficiently detailed to encompass all terrain types. But it does perhaps illustrate the approach advocated here. It also incorporates a subdivision of the vertical weathering profile, since it is emphasised throughout this study that the spatial variation of landscape can only be properly understood in relation to the zonation of weathering profiles and to the outcrop of their different horizons. Birot (1958) applied a similar approach to granite slopes in Brazil and Driscoll, (1964) adopted this method in his description of slope profiles from northern Australia.

These facets may be given more precision by the introduction of local data concerning class limits for slope and texture of parent materials. Information concerning mineralogy or other properties may also be included, and where necessary new facets added to the list. However, once defined for a given occurrence they may be readily combined to describe unit landforms, and these can be grouped into land or landform systems on the basis of position in the landscape and a repeated pattern of occurrence. Table 13 exemplifies this concept for an erosional landscape.

Such a descriptive framework may be applied both to extensive landsurfaces and to individual hillslopes or escarpments. It can be elaborated to reveal greater detail in the study of particular forms (Thomas 1967) or may be simplified to portray the principal characteristics of terrain as in the *Land Research Reports* of the C.S.I.R.O. in Australia. It is possible to regard areas of common pattern as landform systems. Two such systems are described in table 14 from

TABLE 12

COMMON FACETS OF THE LANDFORM IN THE TROPICS

	Slope Forming Materials	*Characteristic Form*	*Common Declivity (Range)*	*Usual Position in Landscape*
Residual materials	1. Topsoil or Duricrust of laterite profile	planar	0–5°	plateau or mesa summits
	2. Laterite rubble	planar or convex	5–15°	convex summits and hillslopes around 1
	3. Laterite cliffs (breakaways)		up to vertical	cliffs around 1
	4. Mottled clay and pallid zones of laterite profile;	concave	2–12°	footslopes around 1/2
	upper zones of non-lateritic weathering profiles	convex or concave	very variable	may occupy whole terrain
	5. Lower zones of weathering profiles containing rock fragments and cores	convex or concave	very variable	may occupy whole terrain
	6. Corestones and tors (granites) and boulder controlled slopes	irregular or planar	up to 25°	tors as residual hills, but cores also on steep hillslopes especially under forest
Bedrock	7. Angular jointed rock	irregular or planar	up to vertical	steep hillslopes and residual hills
	8. Smooth, bare rock surfaces (crystallines)	convex	up to vertical	domed inselbergs and low rock pavements
	9. Truncated rock surfaces	often concave	2–8°	piedmont footslopes
Translocated materials	10. Hillwash (pedisediment)	gently concave	2–7°	valley sides and lower pediments (see also 2 above)
	11. Gravel deposits (pediment gravel)	gently concave	2–10°	valley sides and lower pediments (see also 2 above)
	12. Colluvial material	variable	5–15°	hillslopes
	13. Alluvial and lacustrine deposits	planar	0–5°	plains and valley floors/terraces
	14. Eolian deposits	variable		often as fossil dunes
	15. Cemented deposits (including some laterites)	planar	0–5°	valley-side benches

TABLE 13

Descriptive framework for the definition of tropical landform systems

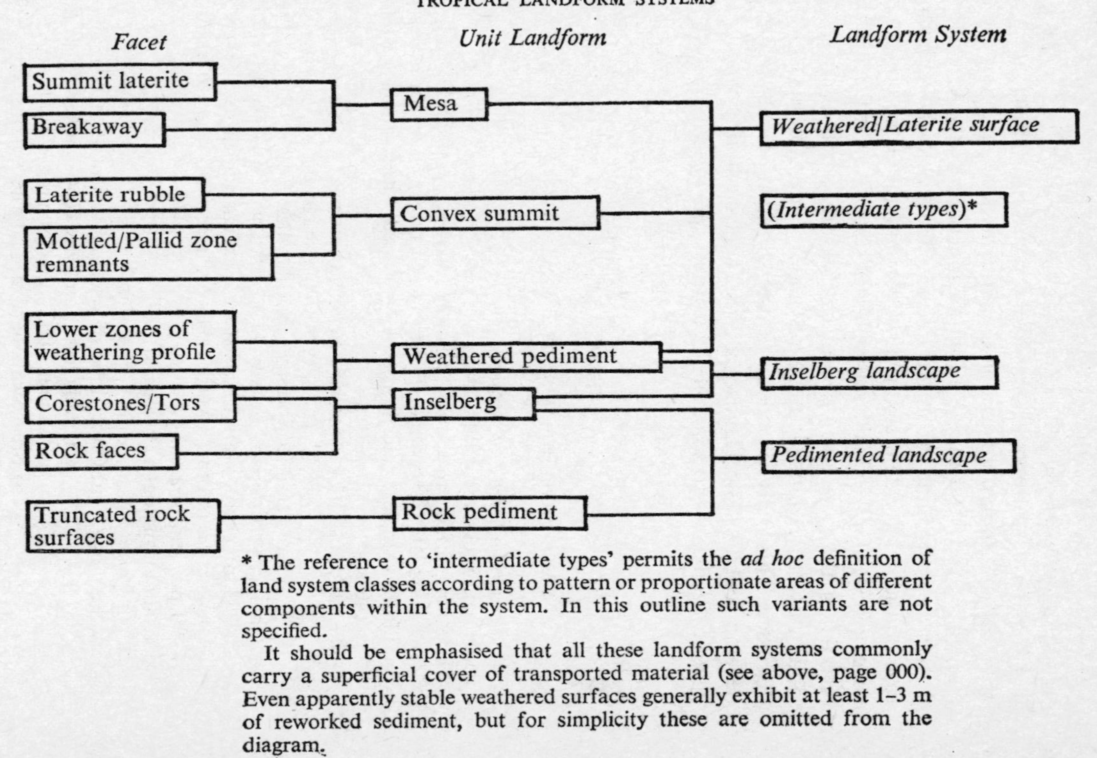

* The reference to 'intermediate types' permits the *ad hoc* definition of land system classes according to pattern or proportionate areas of different components within the system. In this outline such variants are not specified.

It should be emphasised that all these landform systems commonly carry a superficial cover of transported material (see above, page 000). Even apparently stable weathered surfaces generally exhibit at least 1–3 m of reworked sediment, but for simplicity these are omitted from the diagram.

examples of common terrain types in West Africa. It is not suggested that all possible data about such landform systems are included in these examples, and other ways of representing the morphometry of land systems are suggested by Speight (1968, 1970).

The approach adopted here for the description of terrain types may also be applied to complex landforms such as many of the inselberg domes and major hillslopes or escarpments. Conversely, further understanding of the terrain types illustrated above depends to a considerable extent upon an understanding of the nature and development of individual landforms and slope facets within tropical terrain. In the subsequent sections, therefore, some of the major landforms are examined in detail, and theories for their origins discussed. This analysis leads towards a consideration of the overall development of tropical landscapes and to the possibility of assigning genetic significance to some of the terrain types described above.

TABLE 14

DESCRIPTIVE NOTATION FOR TWO LANDFORM SYSTEMS

A. *Dissected weathered landsurface*

Dominant rock type: medium-grained biotite gneiss.

Planation surface remnants: possibly early Cenozoic (African) on summits; late Cenozoic or Pleistocene features within larger valleys.

Landform complex: laterite mesas associated with two-phase stream valleys.

Unit landforms: (1) Laterite mesa with pediment.
(2) Inner valley with active channel.
(3) Rock domes (bornhardts).

Morphometry: drainage density 0·6; mean valley width 1·66 miles (2·6 km); exposed laterite at least 15 per cent of area, and occurring on 45 per cent of transverse valley profiles; rock outcrops less than 0·1 per cent of area.

Detailed morphology:

Unit landforms	*Facets*
(1) Laterite mesa	(a) Crest slope of primary vesicular laterite; slopes generally less than 2 per cent cambered at edges; mean area 1000 square metres.
	(b) Breakaway: hardened laterite cliff or steep rubble.
	(c) Deeply weathered pediment with veneer of fine laterite rubble and concretionary pisolitic gravel; slopes 2–7 per cent, length 200–400 m.

(2) Inner stream valley	(d) Laterite bench of concretionary secondary laterite; slope 2–4 per cent.
	(e) Minor laterite breakaway: hardened laterite forming break of slope.
	(f) Pediment slope; often deeply weathered, but with occasional outcrops, 100–200 m; slope 2–7 per cent.
	(g) Alluvial valley floor.
(3) Rock domes	(h) Convex rocky slopes up to 70 per cent or more; relief 10–50 m.

B. *Dominantly stripped landsurface* (inselberg landscape or etchsurface)
Dominant rock type: variably migmatised, acid gneiss.
Planation surface remnants: none important.
Unit landforms: (1) Domed inselbergs (bornhardts)
(2) Rock pavements
(3) Tors
(4) Open stream valleys without clear evidence of multi-phase incision.

Morphometry: Drainage density 0·99; mean valley width 1·08 miles (1·73 km); exposed rock surfaces at least 7 per cent of area, and occurring on 25·66 per cent of transverse valley profiles.

Detailed Morphology:

Unit Landforms	*Facets*
(1) Bornhardt domes	(a) Many contiguous convex rock surfaces carrying superficial debris; relief exceeds 10 m; bounding slopes generally over 30 per cent.
(2) Rock pavements	(b) Simple convex rock surfaces; relief less than 5 m, slope less than 15 per cent.
(3) Tors (sometimes as landform complexes)	(c) Complexes of contiguous joint blocks.
(4) Stream valleys	(d) Deeply weathered crest slopes; less than 5 per cent.
	(e) Valley flank (pediment?); generally shallow weathering with quartz gravel surface layer; slopes 5–10 per cent; length 0·4 miles (350 m). Interrupted by facets (a), (b), (c).
	(f) Alluvial valley floor.

(See also figure 41).

7 Domed and Boulder Inselbergs (bornhardts and tors)

It was stressed before that definitive characteristics of tors and domes are not sufficiently refined to achieve an unambiguous distinction between the two forms. Characteristically, tors form groups of spheroidally weathered boulders (usually of granite) rooted in bedrock which may be exposed as a basal rock surface or concealed by a waste mantle. Constituent blocks may occasionally reach 8 m in diameter, but 1–3 m are common. On the other hand domed inselbergs typically rise as monoliths which may exceed 300 m in height. In these cases only the summit exhibits pronounced convexity, but smaller domes may be completely hemispherical or may suggest the presence of an incurved underside. Such features, which may be from 5–30 m or more in diameter, cannot be clearly distinguished from tors.

Because of this difficulty, and also because the two forms tend to occur most particularly in granitoid rocks and often coexist in the same local area, hypotheses concerning their development need to be considered alongside. In fact it will be shown that, depending on the evidence adduced, so the origins of tors and domes have been considered either in terms of entirely different mechanisms, or as variants of a single, comprehensive hypothesis.

In such an enquiry it is important to stress that the morphological similarity of some domes and the larger tors does not necessarily imply that a single process results in both spheroidal weathering and sheeting. These two phenomena are generally treated separately in the literature; they are ascribed respectively to chemical and physical mechanisms, and are considered to operate on different scales. If the likely processes at work in each case are carefully examined and their scales of operation defined, it may be possible to consider the possible links between core formation, tor development, and the origins of domes more objectively.

THE GEOGRAPHICAL DISTRIBUTION OF DOMES AND TORS

Domed inselbergs are most strikingly associated with the tropical shields and with granitoid rocks, but they are not restricted to either. King (1957) has emphasised that bornhardts have been 'recorded from every climatic environment on earth', and he concluded that their occurrence is related simply to the presence of hard, massive rock and to the development of sufficient relief. Few generalisations encompass all occurrences. For instance most of the great 'bornhardt fields' are found developed over granites, and they often surmount planation surfaces in Africa, Australia, India and South America. Yet great exfoliation domes occur in the Yosemite Valley (Matthes 1930) and these are clearly unrelated either to the tropics or to plains. Furthermore, the striking domes of the Olga Hills and Ayer's Rock in Australia are built of massive, sedimentary rocks (Ollier and Tuddenham 1961). Although these last are exceptional, it is clear that domical hill forms are not only associated with intrusive or crystalline rocks, and the location of bornhardts in dissected mountain regions such as the Western Cordillera of North America (Cunningham 1971); along the dissected margins of the Brazilian shield (Birot 1968), and as residual hills standing above planate landscapes, also emphasises that they are not restricted to particular topographic positions. According to White (1972) the only areas within which bornhardts are entirely absent are those which experienced intensive glacial erosion. Because these areas in Nozth America and in Fennoscandia are built largely of shield rocks, domes might otherwise be expected in these regions. This contrast in shield morphology has led to an emphasis upon the tropicality of the southern shields, but a case can be made for the glaciation of the northern shields as the more exceptional feature. The occurrence of domical forms over suitable rock formations throughout extra-glacial areas is then very striking.

In a perceptive study of domes in the south-eastern Piedmont of the United States, White (1945) warned against seeking a single hypothesis for all occurrences of the domed form. Comparing the domes of Yosemite and their striking exfoliation patterns, with those of Georgia which display little or no sheeting, he suggested that a principle of 'convergence' might be operative, whereby similar forms have been produced from separate origins via the operation of different processes. This concept has since been formally recognised as the principle of 'equifinality' (Bertallanffy 1950), and has many possible implications for geomorphological studies.

In so far as it is possible to recognise a separate group of tor landforms, their distribution is again extremely widespread, being absent only from some areas of extreme glacial scour. Tors are characteristically associated with regularly jointed granites, but are found also in arkosic sediments and some other crystalline rocks. They are, however, rare among the strongly foliated and banded metamorphic rocks of the shields, and this is in marked contrast to the domed inselbergs which are commonly found among these rocks.

CLASSIFICATION AND LOCATION OF DOMES AND TORS

Notwithstanding the difficulties here alluded to, differences between boulder inselbergs (tors) and domed inselbergs (bornhardts) are generally very marked. Attempts have also been made to differentiate domes according to size and morphology, but these have not been very successful, because they rest less upon observed characteristics of form than upon assumptions about formative processes. Domes range in size from rock pavements and small features less than 10 m across to very large hills greatly exceeding 300 m in height.

Other classifications of both tors and domes rest upon locational criteria, or more accurately their distribution in relation to other landscape features, particularly escarpments. Thus, attempts have been made to classify inselbergs (not always specifically domed features) into zonal and azonal forms. The zonal distribution of inselbergs has been claimed by Passarge (1928), Jessen (1936), Kayser and Obst (1949), King (1948, 1962, 1966), Pugh (1956) and Birot (1958, 1968). Little statistical evidence in support of these claims has been offered, and probably none of these authors would insist that all inselberg forms were zonally distributed. Büdel and Birot make a clear distinction between zonal and azonal features. Birot (1968) distinguishes between the 'inselberg of position' resulting from the retreat of escarpments and the 'inselberg of durability' situated in relation to more resistant rocks, and appearing perhaps towards the centre of a drainage basin or close to a major stream.

Groups of inselbergs have been shown to be located according to major structural alignments in South Australia by Twidale (1964), and similar distributions of inselberg groups occur elsewhere. The role of joints in controlling the isolation of inselbergs has recently been analysed by Hurault (1967), and is illustrated by Twidale (1964) and in this study. This joint control may be responsible equally for inselbergs found close to and far from escarpments, so that zonal distributions may have more to do with the durability of the landforms than with their formation. By no means all domes occupy interfluve sites: they commonly interrupt the general slope of valley sides, in which case the foot-slopes incline *towards* the dome on the upslope side. In the extreme case 'half-domes' emerge from steep escarpments. Savigear (1960) and Birot (1968) have both indicated that, on the wide plains, inselbergs are not arranged systematically with respect to the drainage pattern, and major streams commonly flow around individual domes which form only local hydrographic divides. Although Pugh (1956, 1966) has emphasised the general relationship between inselberg and pediment, which by definition must slope away from the dome, this is not the only, nor in many areas the most common, situation.

Burke and Durotoye (1971) in a recent study based on measurement of distances between residual hills and stream channels of different orders, claim that their evidence supports the view that inselbergs are predominantly situated on interfluves and are by inference surrounded by extensive pediments. The evidence presented shows that, whilst most residuals lie close to 1st and 2nd order streams, they are distant from 5th order streams (Burke and Durotoye 1971, figure 4). This is not surprising, since most bornhardts are large enough to maintain 1st and sometimes 2nd order streams by local runoff and seepage, whilst the infrequency of streams of high order must indicate that a large proportion of all landforms will be distant from them. Unless such figures are analysed in greater detail it is therefore difficult to make deductions from them.

Domes associated with granite intrusions are commonly closely spaced and may be divided only by deep rocky clefts. Such features are found widely in West Africa in association with the Cambrian

Older Granites, as in the Gbenge Hills of Sierra Leone or at Idanre in southern Nigeria, where individual domes rise as much as 600 m above the surrounding plains (figure 30). These hills may be compared with the 'Sugar loaves' near Rio de Janeiro, described by Birot (1958, 1968).

Attempts have also been made to distinguish different types of tor on the basis of position in the landscape. For instance, in the discussion of Linton's (1955) paper on 'The Problem of Tors', King (1958, p. 289) insisted that tors could be classified into two categories: *skyline tors* were stated to occupy high points in the landscape, where 'under favourable circumstances they are circumscribed by smooth, outward sloping aprons that close by the tor may visibly consist of planed bedrock'. Such aprons were described as pediments and the tors were interpreted as residual hills in the final stages of destruction by slope retreat. On the other hand *sub-skyline tors* could occupy valley sides or occur in depressions, where they were not surrounded by rock pediments, and could have arisen from the process of exhumation from a deep regolith as described by Linton (1955).

It is doubtful if this dichotomy can be strictly upheld, but it is interesting that Handley (1952) also made a distinction in terms of location among tors studied in Tanzania. On the 'African' planation surface, tors were found to 'form the highest points in the landscape, and rise steeply from the gently sloping valleys sides' (p. 205). But on younger landsurfaces he found that 'the tors are concentrated in the valley bottoms, or on the valley sides, while the stream divides which are remnants of the African surface are devoid of outcrops and tors except where a residual tor from the African cycle of erosion may still cap the ridge' (p. 205). However, despite the similarity of description, Handley found that 'rock platforms are formed only to a very minor degree, occupying less than 1 per cent of the area between tor and valley bottom, and very often there is no pediment at all'.

Neither Handley (1952) nor later workers in the tropics, such as Ollier (1960), attempted to advance different hypotheses for tors in different locations. Both Pallister (1956) and Ollier (1960) stressed that many 'pediment' slopes bordering tors were in any case cut into weathered materials.

Although tors may be described in terms of location, they do in

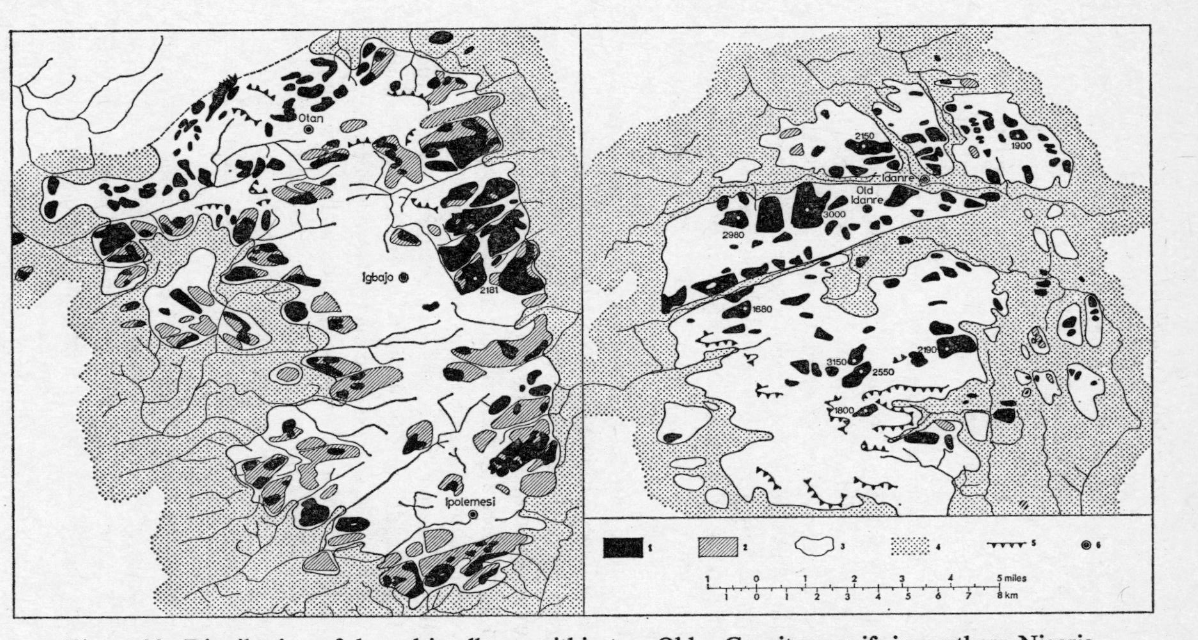

Figure 30 Distribution of domed inselbergs within two Older Granite massifs in southern Nigeria.

Left: Igbajo Plateau. Right: Idanre Hills.

1. Bare rock, domed inselbergs; 2. Domed hills with soil cover; 3. Area of dissected granite country, mostly weathered; 4. Surfaces developed over Basement Complex gneisses; 5. Major breaks of slope; 6. Settlements.

fact occupy many different locations, and the broader view suggests that tors and separated corestones may occupy almost any position in the landscape, and occur in both planate and hilly terrain. Possibly some residual hills occupying summit locations are not really tors comprised of spheroidally weathered blocks, but are kopjes of angular, jointed rock. In some cases these appear to have arisen from the collapse of other more massive forms.

Morphological relationships between domes and tors

Domes and tors often occur together, particularly within granites. But there are several different aspects to this observation. Where they do occur together, there is often a range of block size which gives rise to the confusions in the definition referred to above. Domes may be surrounded by a matrix of loose corestones and occasional tors; tor-like groups of weathered boulders may occur on dome summits, and individual tor groups may have a domical profile. (Plates 15, 17.)

It seems clear that jointing frequencies and the development of different types of joints within a hierarchy must be largely responsible for these observations. In an area where more massive blocks form domes, more closely jointed rock may either be completely weathered or will form corestones and tors. Where sheeting has occurred on dome surfaces, the separated sheets may split up into blocks which become weathered into tor-like forms. The grouping of constituent boulders in tors into domical forms is perhaps the most revealing phenomenon, because it suggests the successive development, first of a dome form and later of smaller spheroidal forms within it. This sequence is clearly illustrated around the margins of the Jos Plateau, where relationships between tor formation and destruction, dome formation and the nature of escarpments is very clear (Thomas 1965a; figure 31).

Such observations as these suggest that tor and dome formation are in certain cases, or at certain stages, linked. They also point to the need for a searching enquiry into the nature of jointing, the role of sheeting, and the development of spheroidal boulders in granitoid rocks.

Areas where tors have not formed or survived in association with domed inselbergs are usually those underlain by gneissic rocks within which small boulders tend to disintegrate along the mineral bands. Where isolated, high bornhardts stand above extensive plains, again few signs of tor development may be present, except perhaps on the dome summits themselves.

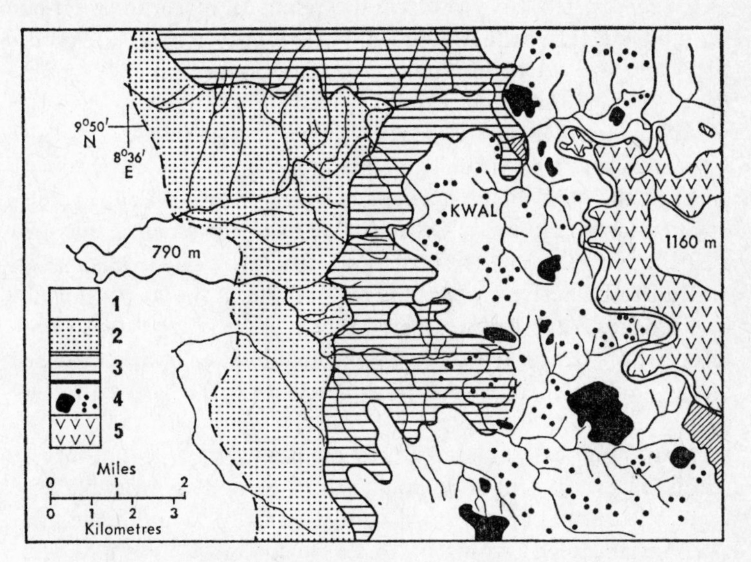

Figure 31 Morphological zones on the western margins of the Jos Plateau, Nigeria.

1. Slightly dissected etchplains; 2. Coalescent alluvial fans – depositional zone; 3. Escarpment zone of stripped half-domes; 4. Partially stripped zone containing domes and tors (tor groups shown as black circles); 5. Basalt lava overlying upper etchplain at 1160 m.

THE PHYSIOGRAPHY OF DOMED INSELBERGS

Before exploring some of the questions posed in the previous section, it is necessary to consider further the character of the larger forms of domed residual. This has two aspects: their detailed relationships with surrounding slopes, and the character of the hills themselves. Many misconceptions exist in both these fields.

Domes and plains

The term 'inselberg' itself suggests that they normally rise abruptly from plains of low relief. This was noted by many early workers such as Bornhardt (1900) and Falconer (1912), and Cotton (1942) who claimed that 'true inselbergs (or bornhardts) rise from plains'. If this is taken literally, it can be demonstrated that not all bornhardts are in fact inselbergs. However, the relationship between hill and plain in the tropics (as indeed anywhere else) may be of critical importance to an understanding of landforming processes and it therefore merits some discussion.

Both King (1948) and Pugh (1956) have insisted that fringing rock pediments surround virtually all inselbergs, and the sharp piedmont angle at the base of the hillslopes has frequently been interpreted in these terms. However, Ollier (1960, p. 47) commented that 'the sharp angle between inselbergs and the surrounding land is not difficult to understand when it is realised that it marks the junction between fresh and very weathered rock'. These observations clearly illustrate differences not only of interpretation but of field evidence, and the hill : plain relationship is in fact not to be resolved in such simple terms.

Many domes occur as features on a single major hillslope or within an escarpment zone, so that they do not rise from plains or possess surrounding slopes declining radially from the hillfoot. Such instances led Birot (1958, p. 24) to comment 'nous n'oublions pas que les pains de sucre bresiliens sont des formes de jeunesse du cycle et non des inselbergs'. This not only makes it necessary to dissociate domes from genetic connotations of the term inselberg but it limits the appropriateness of the term, even for descriptive purposes.

Varying hill : plain relationships amongst domes thus depend upon three main factors: relief development, position in the landscape, and the pattern of surface materials. The widespread occurrence of a 'piedmont angle' is not ubiquitous and depends, by definition, on the existence of a piedmont slope. It is also a misapprehension that rock : regolith transitions are necessarily marked by an abrupt change of slope. In the case of low rock pavements (*ruwares*), the bare rock surface may scarcely break the profile of the plains, but in other cases the marginal slopes of the dome plunge at steep angles (occasionally vertical or incurved) to meet the piedmont at a sharp junction. However, contrary to some descriptions, many domes carry quite large areas of soil cover (Thomas, 1967; figure 32) and this may be continuous from summit to footslopes, across a range of slope angles up to a maximum of 30–35°. Combining these remarks and observations, the following footslope conditions may be recognised:

(1) *Abrupt junctions* usually mark a transition from the rock slope of the dome to regolith, colluvial or alluvial deposits, but sometimes a sharp break of slope exists between two rock faces.

(2) *Facetted junctions* usually occur in the last case above, where a series of continuous, multi-convex rock surfaces form a series decreasing in relief and slope and linking the dome summit to rock pavements of the plains. Facetted slopes also result from accumulations of corestones or talus around the dome.

(3) *Smooth junctions* are usually restricted to places where no sudden change in surface materials takes place at the piedmont angle. This may arise where a concave rock slope descends gradually beneath the deposits of the plains, or where a regolith cover persists across a part of the whole of the dome surface. In these cases a piedmont angle may still be defined, but the transition takes place over a greater length of ground surface.

It is important to realise that all these possibilities may occur around a single dome (Thomas 1967; figure 32), and that they result from and contribute to the asymmetry common to many of these landforms. (Plate 18.) These varying circumstances may also imply that no single mode of slope development takes place around these hills.

Reversal of slope may occur at the hillfoot, where the dome interrupts a major hillslope or where marginal depressions (*bergfüssneiderungen*) occur. These last features have attracted a lot of attention and will be discussed further (see p. 213).

Morphology of domed inselbergs

The most obvious morphological features of bornhardts are their domical form which may only amount to a pronounced summit convexity, steep flanking slopes which are dominantly of bare rock, and clear differentiation from surrounding terrain. In addition, these features may be associated with curved rock plates or sheets. Some writers have regarded this phenomenon as fundamental and described the hills as 'exfoliation domes'. The convexity of profile is occasionally accompanied by circularity of plan (Hack 1966; Cunningham 1971), but in most cases bornhardts exhibit rectilinear

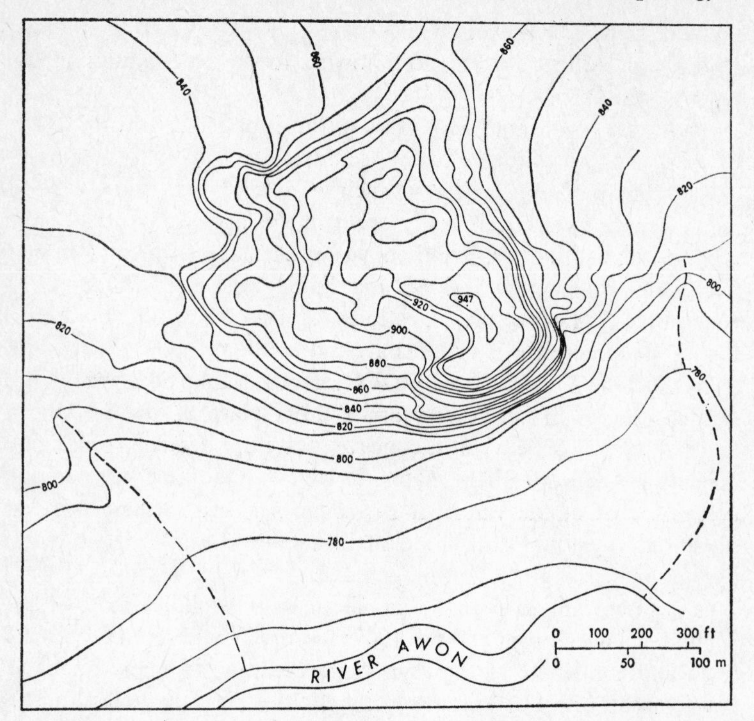

Figure 32 Morphology of a gneiss inselberg in the savanna zone, near Oyo, southern Nigeria.

A. Contour map

plans, apparently controlled by fracture patterns (Twidale 1964, 1971; Hurault 1967; Thomas 1967). King (1948), however, denied that the actual boundaries of the hills were joint controlled. Whilst some domes in granites are nearly symmetrical, asymmetry is common. This often results from internal structural patterns in metamorphic rocks.

Summit convexity provides the main diagnostic characteristic of these forms, but lower slopes may be convex, constant or concave. (Plate 18.) Concavity has been attributed to water flow by King (1948, 1953), but radial dispersion of runoff, and lack of debris able to erode the bare rock surfaces makes this simple explanation unlikely. Various forms of basal weathering attack may be responsible as discussed here.

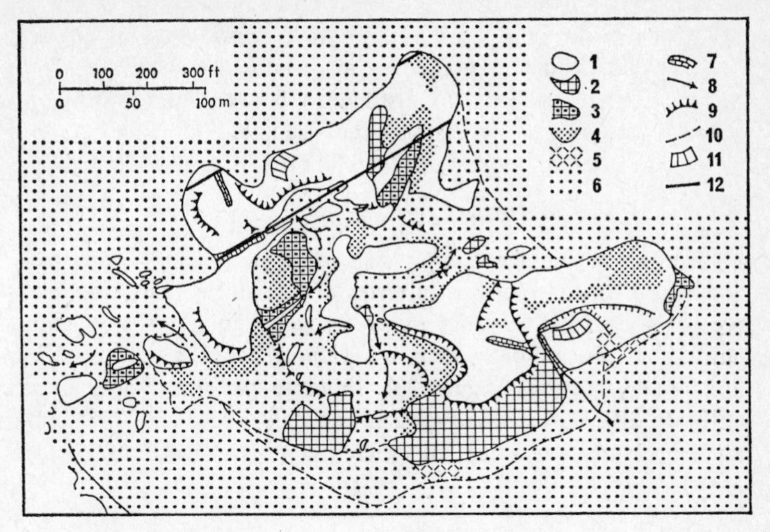

B. Morphological map

1. Smooth, bare rock surfaces; 2. Jointed and broken rock surfaces; 3. Partially rotted and fractured rock; 4. Debris streams of small, broken and etched rock fragments; 5. Talus blocks; 6. Soil covered surfaces; 7. Rock gullies; 8. Permanent channels in soil covered surfaces; 9. Convex breaks of slope; 10. Concave breaks of slope (these also occur at the junction between rock and soil surfaces): 11. Exfoliation slabs; 12. Dolerite dykes.

The summit morphology of larger domes may be complex and they may exhibit several cupola-like culminations, almost planar bevelling, and in some cases a substantial regolith cover. Laterite has been identified on some dome summits (Falconer 1911; Eden 1971), and lakes may occur in depressions.

The morphology of tors

Boulder inselbergs or tors are generally comprised of one or more spheroidal blocks, and ambiguity in relation to dome only occurs if these are very large and only partly exposed. Most tors are less than 30 m in total height and constituent blocks are seldom more than 8 m in diameter. Some are clearly defined by a rectilinear jointing lattice, but others appear related to the weathering of curvilinear sheets. Tor profiles may be highly irregular, but some conform to a domical outline, and where constituent blocks are little weathered merge gradually into true domes. (Plate 15.) In many cases it is difficult to distinguish between tors founded in bedrock and groups of corestones that are not (Linton 1955).

INSELBERGS AND JOINTING

It is clear from the foregoing discussion that tors and domes may commonly be distinguished in terms of size and massiveness. Tors are generally subdivided by quasi-horizontal joint systems; bornhardts are not, although their morphology may be influenced by horizontal structures (figure 33). Fracture patterns clearly affect both the distribution of inselbergs (Twidale, 1964) and their morphological characteristics. By inference, domed inselbergs would

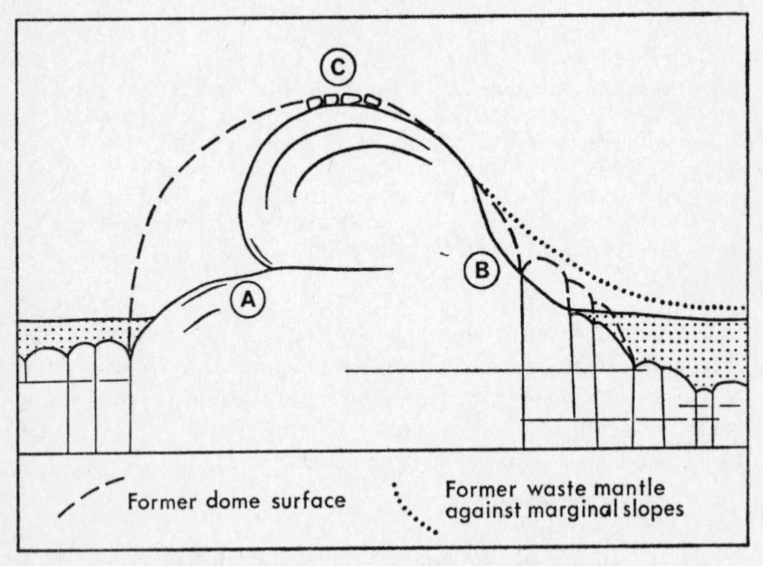

Figure 33 Some common morphological features of domed inselbergs.

A. Incurved face and rock apron developed along sub-horizontal line of weakness; may show sympathetic sheeting and/or taffoni developments; B. Concave rock surface, postulated as inherited from a former waste mantle; C. Tor blocks from fractured exfoliation sheet.

appear to occur in more or less undivided masses of rock, and therefore in areas of infrequent jointing. But this simple statement must be carefully examined, because on the one hand tors may occur in domical groups, while on the other hand domes may become subdivided or may even collapse as the result of the opening of plane joints. Several interrelated problems may be identified in this situation: these are particularly concerned with the origins of different joint sets and with the role of these fractures in the development of forms which might otherwise be seen mainly as a result of a particular denudation system.

Jointing in intrusive rocks such as granites may result from stresses set up during cooling and crystallisation (see Balk 1937); from regional tectonic stresses applied to the rock since its emplacement, and to past and present relief development. In granites the upper parts of intrusions may be more closely jointed than the deeper masses (Balk 1937; Cunningham 1971; Twidale 1971), so that depth of erosion may influence the nature of granite relief. The rectilinear jointing appears to define the size of rock masses from which later landforms have been developed, but Twidale (1964, 1971) has emphasised that many rock masses exist under strong lateral as well as vertical stresses, and that domes may be located in areas of high lateral compression. Once formed, any landform is affected by boundary conditions and both theoretical (Gerber and Schiedegger 1969) and statistical studies (Chapman and Rioux 1958) have indicated how these may contribute to the opening of latent weaknesses into joints. But perhaps the most controversial topic surrounding the development of domed inselbergs is that of sheeting.

Sheeting of bornhardt domes

Many authors implicitly, if not explicitly, attribute the distinctive domed shape of bornhardts primarily to the development of sheeting joints. On most domes separated sheets of unweathered rock can be seen in various stages of disintegration into smaller joint blocks. This sheeting is by no means confined to tropical latitudes, indeed much of the best developed sheeting has been recorded in formerly glaciated, highland regions such as the Yosemite Valley (Matthes 1930) and in Greenland (Oen 1965). Concerning the high Sierras of North America, Gilbert, as early as 1904, recorded that 'the granite is divided into curved plates or sheets which wrap around the topographic forms. The removal of one discloses another, and the domes seem at the surface to be composed, like an onion, of en-wrapping layers' (p. 29).

The sheets vary in thickness from a few centimetres to several metres, though most fall within the range between 0·5–1·5 metres. A number of authors have noted that the sheets become thicker with depth from the outer rock surface (Dale 1923; Jahns 1943; Chapman and Rioux 1958; Terzaghi 1962; Twidale 1964). In most recorded instances of sheeting the phenomenon dies out with depth. Gilbert (1904) considered 30 m to be a common limit but Jahns (1943) cited an isolated occurrence at 100 metres. In Brazil, Birot (1958) found that sheeting did not extend to the centre of the domes, and that four or five superposed layers were normal.

A comparison of observations from glaciated and tropical regions indicates that sheets are more numerous in the former. White (1945, p. 216) commented with regard to domes in Georgia and North Carolina that: 'those (domes) which have developed from granite and gneiss of the south east do not show enough evidence of exfoliation to justify the assumption that it has played a dominant or even important part in their formation.' This observation is supported by the work of Hurault (1967), based on studies in Guyana and Gabon.

On tropical inselbergs most sheets are convex, but concave sheeting is known, especially from glacial troughs in high latitude regions. Most sheets are in fact lenticular with their thick and thin parts alternating. According to Jahns (1943, p. 73), 'with increasing depth the sheets become progressively thicker, more nearly horizontal and less irregular'.

Weathering along the partings between sheets is usually absent, but it is common for the separated sheets to break up along radial or 'slope' joints and circumferential or 'contour' joints (Chapman 1958). On steep slopes the blocks fall or slide to form a talus, but on gentle slopes they remain and slowly weather to more irregular shapes. In some cases the separation of sheets appears to have occurred beneath a regolith cover, and the resulting blocks have become spheroidally weathered into cores or tors.

Relationships between sheeting and topography vary. A general parallelism between sheets and surface form is almost axiomatic, but divergence between the two is sometimes highly significant. It is important also to recognise that not all fractures parallel to the landform are necessarily due to sheeting: some may be a part of earlier joint systems (Harland 1957; Cunningham 1971).

It is interesting that, while in some cases sheeting has developed parallel to very recent topographic forms, as noted on massive sandstones in Colorado for instance (Bradley 1963); in other areas the sheets are truncated by such recent erosion. In many glaciated areas sheets have been truncated by late glacial or post-glacial erosion. This was found in Acadia National Park, Maine (Chapman and Rioux 1958); and on Dartmoor, Waters (1952) found that sheets running parallel to an older landsurface had been truncated by recent stream erosion. Birot (1958, 1968) has shown how plane jointing may intersect sheeting on the flanks of bornhardts.

The lack of consistency among these records has contributed to disagreement concerning the rate at which sheeting occurs. Matthes (1930) considered the process to be very slow, and his views are followed by Chapman (1958) and Keislinger (1960), who concluded from observations in Norway that the spalling of sheets requires a lengthy geological time span. On the other hand Bradley (1963) concluded that 'expansion and rupture occur very soon after pressure is released and very slowly or not at all thereafter'. In fact the disagreement probably concerns not the rate at which sheets may be shed, but whether the process is continuous, and can lead to progressive reduction of relief. The weight of evidence appears to support Bradley's view of the phenomenon as will be shown.

Other aspects of sheeting have some importance to the present discussion. The fractures do not follow internal structures in rocks, and the phenomenon is not confined to crystallines. Relationships with other joint sets probably depend on the age and origins of each. Many domed inselbergs possess several differently centred sheeting systems, each of which terminates at a master plane joint, and Twidale (1964) claims that sheets often plunge steeply close to the margins of major joint blocks. Chapman (1958), however, recorded plane joints truncated by sheets, and the intersection of rectilinear and curvilinear joints is recorded by Bradley (1963) and Oen (1865).

Although sheeting is usually associated with large rock masses it can affect spheroids (cores and tors) as small as 8 m in diameter, and under artificial conditions, as in mine tunnels, curvilinear partings can be produced with diameters less than 1 metre. The possible association of sheeting with corestones, and therefore with concepts of spheroidal weathering, further obscures the distinction between the two major forms of inselberg.

The causes of sheeting

Sheeting appears to result from the relaxation of residual stresses in rocks. It is a spontaneous process of dilation, taking place towards surfaces along which confining pressure is being, or has been, reduced. However, the sources and the patterns of stress fields in rocks still pose unanswered questions, and the manner and rate of relaxation remain matters of discussion. The simplest explanation of sheeting follows the work of Gilbert (1904), who accounted for it in terms of unloading. Granites formed from magmas at depth would store compressive stresses sufficient to bring about actual expansion once the superincumbent rock is removed. Deep burial may also affect sedimentary rocks, so that this hypothesis can be applied to most known occurrences of sheeting. Oen (1965) considered that density differences between late tectonic granites and country rock could cause compression of the upper part of the intrusion and lead to expansion and sheeting, following upon uplift and drainage incision.

Stress fields in rocks arise, however, from other causes, including original crystallisation, recrystallisation during metamorphism and recent tectonism. Twidale (1964, p. 104) suggests that radial compression may affect rock masses such that 'an upward release of pressure would result in the arching of the rock in a few geometrically arranged localities'. Massive bornhardts could thus be associated with locations in which rock has formerly been under high radial compression. The subject of stress fields is complex and they clearly operate at greatly varying scales (Scheidegger 1970). Although Twidale (1971) has argued against the unloading hypothesis it cannot be discounted. It does not however appear to account for all observed features of sheeting.

Sheeting and landform development

Because of the close association of sheeting with surface form, there is a danger of circular argument in the search for cause and effect relationships between sheeting and domed inselbergs. This problem can be approached in terms of the manner in which compressive stress appears to be released in massive rocks.

Many observations emphasise the parallelism of the sheeting with old landsurfaces. The indication that rapid incision may truncate sheets is possibly important to the present discussion, but the development of 'new' slopes on old forms can in some cases lead to renewed sheeting. This was noted in sandstones by Bradley (1963) and can be seen in rock gullies developed along plane joints traversing bornhardts.

Sheeting joints are also suggested from boreholes penetrating below deep regoliths, and it is possible for stress relaxation to take place with respect to weathering patterns. Decomposition of crystalline rocks leads to a loss of some constituents, an increase in voids and a decrease in density of the regolith. It is therefore reasonable to suggest that, if weathering penetrates deeply along master joints or shatter zones, sheeting will occur concurrently within the intervening rock mass (Thomas 1965a). If this process is widespread, then a relationship between sheeting and weathering patterns may develop independently of the detailed form of the landsurface. The presence of spheroidally weathered blocks on dome summits, underlain by unweathered sheets, suggests that some sheeting may take place in this manner.

According to Jahns (1943, p. 84), 'sheeting broadly controls topography when denudation is relatively slow and long continued but is reoriented when strongly concentrated erosion by water or ice creates relatively rapid topographic changes'. This comment supposes sheeting to be a slow and continuous process, and Mabbutt (1952, p. 91) has suggested that, 'reduction of domes takes place by the shedding of successive rock plates'. Matthes (1930) also considered sheeting to be responsible for the convexity of summits and for the wholesale reduction of rock masses.

But if Bradley (1963) is correct to suggest that once stress is released during an early stage of landform development, little or no further exfoliation will occur, then the process is self limiting and unlikely to lead to the destruction of large domes. There is a lot of evidence to support this viewpoint, particularly the relative rarity of sheets on tropical domes which may reasonably be assumed to have developed slowly over long periods of denudation. Furthermore, the persistence of fractured sheets on bornhardt summits, and the general absence of significant talus accumulations below many domes points to the cessation of the sheeting process.

According to Gerber and Scheidegger (1969, p. 407) 'all processes, whereby unstable decay-prone forms are transformed into more stable ones, occur very rapidly. If these decay processes have run their course, a more stable form remains which is able to subsist for a relatively long time and decays only when weathering and erosion have produced new unstable forms.' This principle, if sound, may account for the long persistence of very ancient domes on the tropical shields (Thomas 1965a).

The contribution of sheeting to the domed form of bornhardts is not defined by these arguments; both weathering and sheeting may contribute, and this recalls the comment of White (1945, p. 282) that the domes form 'is the expected form wherever non-jointed homogeneous rocks are subjected to the attack of any non-directional agency of denudation. The particular agency would be determined by local climate but the resulting landforms would be essentially similar'. This is a classical dilemma in geomorphology, when the slowness of the denudational process defeats any attempt to measure its long-term effect upon landform. In the tropics the interaction of the physical process of stress relaxation and chemical weathering appears to be the central question to be considered, and this must include enquiry into other forms of jointing and weakness in rocks.

Possibly this problem is most clearly focussed by the problem of corestone formation in regoliths and the appearance of surface tors. On large domes the sheeting process often appears to be almost entirely physical, while the formation of small cores in granite regoliths is commonly attributed to the operation of chemical weathering. The difficulty in distinguishing the larger cores from domes suggests that the two processes must interact, and the absence of cores from many granite regoliths may also be relevant.

THE PROBLEM OF SPHEROIDAL WEATHERING

In a recent review of spheroidal weathering Ollier (1971) has stressed the importance of distinguishing this phenomenon from other forms of exfoliation such as sheeting and surface scaling (lamination). Because spheroidal weathering is associated with the presence of a regolith, and affects the entire boulder including its underside, and because it is a process found equally in extrusive rocks such as basalt that have not been subject to deep burial, it is necessary to consider the process quite separately from sheeting even if the two interact. On the other hand some well jointed granites do not appear to form corestones and the reasons for this are not fully understood.

Spheroidal weathering appears to result from the encroachment of chemical weathering upon cuboid joint blocks, but the mechanism is not clear. The process gives rise to concentric rings of progressively decomposed rock passing finally into a structureless regolith (Nossin 1967). It is these exfoliation effects which recall sheeting, but they occur on a different scale, being only 10^{-2} m in thickness.

Corestones seem to form in uniform and well jointed rocks that have not had their fabric weakened by deep seated alteration such as biotite oxidation (Eggler, Larson and Bradley 1969), or by intensive shattering during uplift. Many rocks are penetrated by 'microfissures' (Birot 1962) or 'microcracks' (Bisdom 1967). Chapman (1958, p. 55) described these as 'potential joints' which are 'highly discontinuous faces or thin zones of weaker material . . . represented by cracked grains, cleared grains, disjointed grain boundaries, tiny faults, and layers of tiny fluid inclusions'. According to Birot (1962) these microfissures largely control porosity in crystalline rocks and are consequently of great importance to weathering processes. He attributes these microscopic fractures to stresses applied during different phases of crystallisation or re-crystallisation; to the sum of tectonic influences and perhaps to hydrothermal action and even the influence of atmospheric weathering itself. Measurement of the incidence of fissuration in quartz from near Rio de Janeiro confirmed that there is general relationship between the density of visible joints and of microfissures, and also close correspondence between the fissuration of quartz and landforms in that area. On dome faces more than 50 per cent of the quartz was found intact, but below cols this figure dropped to 20–30 per cent.

The occurrence of these microcracks may be critical to accounts of spheroidal weathering and corestone formation, and for two reasons. First, rocks penetrated by a dense network of microcracks may break down to gruss rather than form cores at all, but secondly, because studies have shown that the inward progress of weathering which forms cores from original joint blocks may be guided by systems of microcracks (Bisdom 1967). Bisdom described linear, structural microcracks less than 10 μm in width, with a transition to more sinuous and wider cracks (30–45 μm) which follow upon weathering. Initially, the cracks appeared to converge normal to the core, but in the later stages of decomposition they become parallel to the core surface, and scales and flakes are formed along macro-cracks more than 1000 μm in width.

These fissure systems cannot alone account for the progress of spheroidal weathering, and Bisdom suggests that they may be related to the diffusion mechanism discussed by Augustithis and Otterman (1966). These authors were concerned with the processes leading to concentric colour banding associated with corestone development. These bands were interpreted as *Liesegang rings* resulting from rhythmic precipitation from diffusion of super-saturated solutions through intergranular spaces in the rocks. They found that the brown bands were enriched in iron and calcium, but depleted of other elements. Iron content in these bands was nearly four times that found in the white layers. A diffusion process leading to progressive depletion of soluble constituents in the rock and producing mineral banding due to periodic wetting and drying and rhythmic precipitation could, in conjunction with the observations on microcracks, account for the gradual formation and final destruction of corestones in jointed rocks.

Such observations may be of considerable help in discussion of the problems surrounding dome, core and gruss formation. It may be suggested that domes form in rocks that are substantially free not only of open joints but also of microcracks, while deep gruss formation will occur, where the rock is both closely jointed and also weakened throughout by mineral alteration and/or microcrack systems. Corestones appear to form in intermediate situations where the weathering front advances gradually into jointed rock with the exploitation and/or development of microfissures. Bisdom (1967) points out that concentric scales are released from cores in a zone within which the rock is no longer likely to store residual stresses, so that the process is not akin to sheeting. However, large cores also exhibit sheeting fractures, so that the two distinct processes interact at the upper and lower ends of the scales at which they separately operate.

THE DEVELOPMENT OF TORS IN WARM CLIMATES

From the accounts of weathering processes, profile development and core formation it is a short logical step to suggest that boulder inselbergs or tors are no more than exposed corestones which rest on foundations of solid rock. This reasoning was implicitly or explicitly adopted by many early writers on granite landforms (T. R. Jones 1859; J. C. Branner 1896; R. Chalmers 1898). Falconer (1911, p. 247) observing a deep and lateritised weathering profile at Kano in Nigeria, commented that the hill 'although deeply decomposed, still preserves in its lower part detached boulders or cores of unweathered rock. If the subsequent erosion had continued until the weathered material had been entirely removed, the flattened hill would have been replaced by a typical kopje* of loose boulders resting on a smooth and rounded surface of rock below.' C. A. Cotton (1917) applied a similar reasoning to the evolution of schist tors in New Zealand. Many others have formulated similar hypotheses including Handley (1952), and Linton (1955) who made a wider public aware of the implications of such an hypothesis, not only for tropical and subtropical latitudes but also for landform study in high latitudes.

Linton's (1955) analysis has often been described as a 'two stage hypothesis' of tor formation in which prior deep weathering is succeeded by a period of intensified erosional stripping. This formal view appeared appropriate to south-west England where Tertiary climates are known to have been warm, and Pleistocene climates periodically periglacial. It also made the assumption that contemporary weathering processes would be slow.

These constrictions need not apply to tropical climates, and in fact may be less important to higher latitudes than is often supposed. For a wide range of seasonal and humid climates in middle and low latitudes rock decay may be regarded as a continuous process, leading to deep decomposition in silicate rocks whenever and wherever the rate of surface erosion is slow. Periodic acceleration of surface denudation may, however, overtake the process of weathering to exhume corestones as tors. Such circumstances can result from

* This confusion between tor and kopje is common in the literature. Both terms arise from colloquial usage and neither is therefore precise, but in this study kopjes are regarded as features of sub-aerial joint collapse ('castle kopjes').

increasing aridity of climate, or uplift and tilting of the land-mass.

Widespread stripping to expose tors thus results from major disturbance of the denudation system and this frequently follows climatic change. However, along escarpments, corestones and occasional tors may be exposed during systematic slope development, apparently without the intervention of such disturbances. This takes place beneath a woodland or forest cover, but under more open communities slopes over 30° are generally dominated by rock exposures.

However, independent evidence for climatic change exists in many tropical areas, and particularly in the savannas where tors are most commonly encountered. This suggests that most tors have passed through periods of stability and instability in the landscape, or may have resulted from such changes. Tors found within a zone of pronounced stripping around the Jos Plateau in Nigeria (figure 31) are found to occur above an escarpment, at the foot of which have accumulated large alluvial fans which have been interpreted in terms of Pleistocene climatic oscillations (Bond 1949).

Confusion concerning the development of tors tends to arise from two causes: first, because there is a tendency to view their evolution in terms of a formal two stage hypothesis and to abandon the discussion once the rock landform has been exposed, and secondly, because the 'age' of tors can seldom be conclusively proven. In south-west England Linton (1955) deduced that stripping would have occurred during Pleistocene cold climates, and it is clear that many tor groups on Dartmoor are encircled by boulder fields and head deposits. In contrast, tors on the Monaro tableland of New South Wales were considered by Browne (1964) to have survived from the Oligocene, because basalt lavas thought to be of this age had manifestly flowed around them. On the other hand cores ejected from shallow regoliths on steep hillslopes may be considered contemporary (plate 13).

There need be no contradiction among these observations, but it is necessary to regard core formation within granite regoliths as continuous, and to recognise that instability of the landsurface leading to tor exposure is unlikely to have been restricted to a single period. Tors may therefore be very old or quite recent in origin. On the other hand they are often temporary forms in the landscape, experiencing a cycle of growth and decay.

In well jointed granites, tors once exhumed are exposed to a sub-aerial regime which leads to rapid drying of rock surfaces after rain, and therefore to a retardation of weathering processes. The smooth spheroids commonly become diversified by weathering pits that form where water lingers on the rock surface, or they may undergo fracture or splitting due to the tensional forces acting on the boulders. However, the underlying foundations of the tor become subject to continued decay within the vadose zone. As these foundations rot and the regolith is slowly removed by downslope processes, the former tor settles or collapses into groups of separated corestones. This process of tor decay can be identified from the disposition of boulder groups and from sections (plates 15, 16) in which rotting beneath the tor group may be inferred. In zones of dissection, such tors appear to form, decay and re-form at a lower level in a systematic manner, possibly in response to oscillatory changes in the denudation system (figure 34 and Godard 1966).

In some cases the tors overlie dome summits or at least intact and massive sheets of granite. These foundations seldom decay, but if subsequent arguments are followed the dome may be gradually exhumed carrying the tor as a crown of boulders on its upper slopes.

Since corestones and tors may become exposed under a varity of conditions, it is important to consider the landform system within which the tors are found. Thus, tors which form summit groups in basined or planate landscapes are often a result of remote events; on the other hand tors exposed on steeper slopes or isolated on summits and flanks of dissected convex compartments may be recent and currently forming. The tor as such is therefore probably a poor indicator of landscape history (Thomas 1974).

The suggestion of King (1958) that 'skyline tors' were produced by sub-aerial scarp retreat and not by exhumation, depends upon the acceptance of the term tor to describe features which I should personally prefer to describe as kopjes. Nevertheless, there appears no reason to suppose that tors may not undergo some retreat of marginal slopes after exhumation, so that the two concepts need not be seen in opposition. Many tors, possibly in a similar way to domes, become isolated during slope retreat. But they become isolated because they are less weathered and more resistant than the surrounding matrix of shattered or weathered granite. The degree of decay in the surrounding material need not be great, but if they

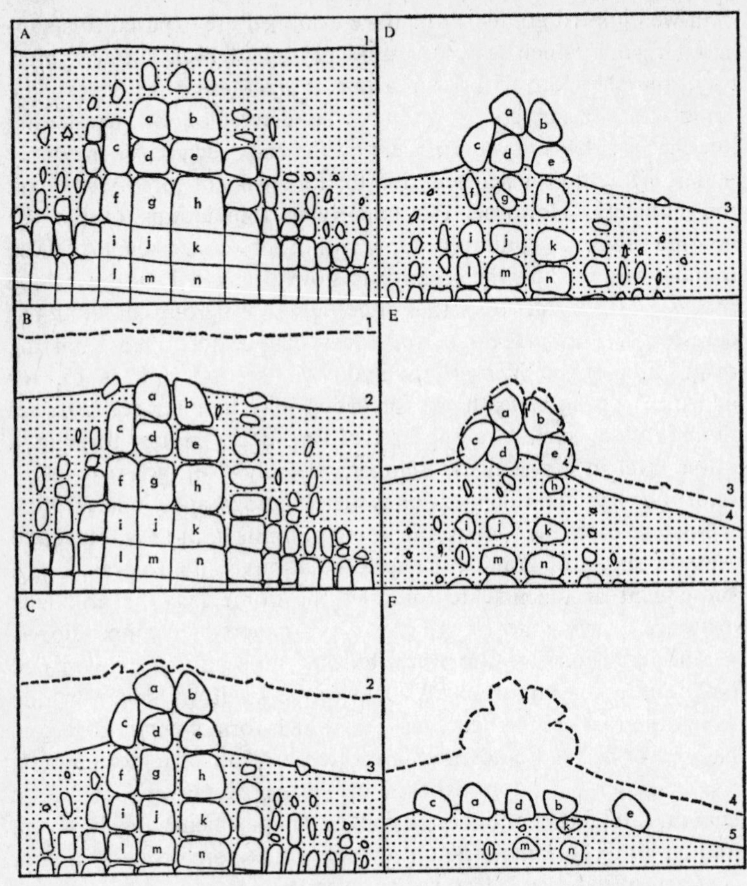

Figure 34 Schematic representation of the development and decay of granite tors (after Linton 1955, Thomas 1965).

A Differentially weathered granite (joint blocks are indicated a–n; B emergence of joint blocks a, b as a result of ground surface lowering from 1 to 2; C tor comprised of blocks a–e exhumed from regolith following further surface lowering from 2 to 3; D foundations of tor become rotted as blocks f–n remain subject to continued groundwater weathering; E tor subsides as ground surface is lowered from 3 to 4; F tor becomes reduced to a group of residual corestones as further lowering of surface and continued groundwater weathering take place (3 to 4 to 5).

arise due to differences in jointing frequency, then the more closely jointed rock is usually found to be in a more advanced state of weathering.

This line of reasoning only becomes subject to major qualification where frost intervenes within the denudation system. Such conditions are specifically excluded from this study, but they do obtain in many localities where tor development has been discussed (Linton 1955; Demek 1964). Because external form may be a poor guide to genesis, the relative roles of frost weathering and chemical decay in the formation of such polygenetic features are difficult to determine. Nevertheless, the widespread occurrence of deep weathering profiles throughout the extra-glacial zone (see chapter 11), and particularly within areas of tor forming granites, provides strong circumstantial evidence for exhumation as the primary process leading to tor formation, even if periglacial mass movement and subsequent frost shattering were closely associated with this development and with subsequent changes in the external form of the tors.

The occurrence of hydrothermal alteration in granites, for instance on Dartmoor (Palmer and Neilson 1962; Eden and Green 1971), although of local importance, does not fundmentally affect the validity of either the concept of widespread deep weathering of granites or the subsequent exhumation of tors. It is, however, a factor which must be recognised as important within granite areas, and particularly so since diagnostic features of hydrothermal alteration are not always present or understood. Deep weathering of the rock may follow slight alteration from deep seated causes (Eggler, Larson and Bradley 1969), and thus two decay systems of separate origins and widely differing in period of operation may interact to give rise to deeply rotted granites. To this must be added the observation by Barbier (1967), quoted by Twidale (1971), of Precambrian weathering of granite near Tassili in the Sahara.

It thus follows from these observations that neither the alteration of the granite nor the exposure of the tor landforms can be restricted in time, except in so far as corroborative evidence from datable deposits is available.

THE ORIGINS AND EVOLUTION OF DOMED INSELBERGS

The frequent association of tors with domes in the landscape suggests that both forms may have a common formational history.

Yet, although a measure of agreement exists concerning the development of tors in warm climates, a vigorous debate has surrounded the origins of bornhardts. Thus, although in 1962 Cotton felt able to state that 'it is becoming increasingly clear that inselbergs are residuals of rocks that are particularly resistant to high-temperature chemical weathering and that they emerge when the regolith over the rocks more susceptible to such weathering is eroded' (p. 279), King (1966, p. 98) considered that such an hypothesis 'fails utterly to account for the great bornhardt fields of Africa or South America with their many hundreds of feet of relief'. It is necessary to trace the development of the divergent ideas as to their origins and to examine them in relation to the field evidence.

Although some early writers favoured the development of bornhardts by aeolian or marine action, neither of which need be entertained today, Branner in 1896 recognised the importance of deep weathering in the formation of the domes near Rio de Janeiro in Brazil. But we owe to Falconer, who worked in Nigeria in the early decades of this century, the first explicit account of bornhardt development by exhumation: 'A plane surface of granite and gneiss subjected to long continued weathering at base level would be decomposed to unequal depths, mainly according to the composition and texture of the various rocks. When elevation and erosion ensued the weathered crust would be removed, and an irregular surface, would be produced from which the more resistant rocks would project' (1911, p. 246). He regarded the domed form as the most stable form, resisting further action by agencies of denudation. Falconer continued: 'a repetition of these conditions of formation would give rise to the accentuation of earlier domes and kopjes and the formation of others of lower level' (p. 246). A similar account of these landforms was also given from Moçambique by Holmes and Wray in 1913, and support for such ideas came later from Bailey Willis working in East Africa (1936). He thought that rock decay was the 'determining process in the isolation of an inselberg' (p. 120), and listed four prerequisites for their formation: 'one, a terrain composed chiefly of gneiss or schist, intruded by granite and traversed by veins of aplite and quartz; two, vertical or steeply dipping schistosity and jointing, which in general facilitates the decomposition of the rock and which serves to give precipitous faces to more massive and more quartzose bodies; three, a climate characterised by warmth and humidity, favourable to abundant vegetation and the resultant processes of rock decay; four, notable uplift which in the usual case will be found to have progressed . . . with marked variations in the rate of elevation' (p. 120). Willis recognised the need for repeated uplifts to achieve heights exceeding 30 m or so, but thought that, 'given a succession of them the altitude which an inselberg may eventually attain above the surrounding plain will depend on the endurance of its summit against spalling and the resistance of the mass in general to decay along joint planes'.

Similar ideas of dome formation were later advanced by Rougerie (1955), working in the Ivory Coast; Büdel (1957), from studies in Surinam, and Birot (1958) from further work in Brazil. Ollier (1959, 1960) subsequently applied these hypotheses to inselbergs in Uganda, where he recognised that the intervening plains were deeply weathered. In 1961 Mabbutt demonstrated the operation of the stripping process in the exhumation of domes in Western Australia where he concluded that 'granite domes have been and are being exhumed from the weathering profile. They are regarded as semi-ellipsoidal kernels of unweathered rock due to a form of structural compartmentation which has controlled the weathering front' (p. 113). In this case only small domes, less than 30 m in height, were involved and a single phase of stripping invoked. Similar arguments have been advanced by Twidale (1964), Thomas 1965a, 1966a) and Doornkamp (1968).

The principal problems encountered by this exhumation hypothesis concern the polyphase development necessary to explain high domes, and the mechanisms of stripping, particularly within areas of rain forest. The question of stripping was considered at length by Birot (1958), who considered that mass movement could bring this about under conditions of high relief, especially when the basal surface of weathering was particularly abrupt. But he also invoked an arid phase to account for widespread evidence of rapid erosion in the form of mud flows and cones of dejection in the steeply dissected country near Rio de Janeiro. Similarly Bakker and Levelt (1964) required savanna rather than forested conditions for the maintenance of bare-rock slopes in the tropics. Rougerie (1955) and Hurault (1967) have both commented that over much of the forest zone the topography is developed on the mantle. But there is also ample observation to confirm the presence of bornhardts (or domed inselbergs) within, the rain forest areas. (Plate 19.) The question which will be discussed subsequently concerns whether they were developed as a result of some former, more arid, climatic phase.

A number of writers have considered the exhumation theory capable only of explaining the smaller inselbergs, thus forming the basis of a division into different types discussed above. Bakker (1960) considered that the absence of a regolith several hundreds of feet thick between domes argues against exhumation except for small inselbergs and during the initial development of larger ones. Berry and Ruxton (1959) were also hesitant on this issue and confined their comments on stripping to the possible formation of small domes.

Lester King in 1948 argued for bornhardt formation as a result of pediplanation and slope retreat within structurally controlled compartments, and with little variation he has held to this view (King 1957, 1962, 1966). According to this hypothesis domes develop when, 'after the incision of rejuvenating streams, the lateral scarps or valley walls standing steeply in the hard, homogeneous rocks, retreat parallel to themselves, maintaining that steepness of flank for which bornhardts are notorious. The retreat is sometimes by spalling, when the base of the bornhardt slope is littered with debris, and sometimes by chemical weathering' (1948, p. 85). This scarp retreat leaves a surrounding pediment at the foot, and 'the occurrence of scarp and pediment, by whatever processes these are formed is typical and is a function wholly of the new cycle, and cannot conceivably be related to any form of weathering beneath an older landscape' (1966, p. 98). He further criticised the exhumation hypothesis on the grounds that bornhardts commonly greatly

---

A, B, C, indicate massive granite blocks, divided by major fractures followed by drainage lines. Stages in the development of the inselbergs are shown: A_1 B_1 C_1 landsurface formed on regolith (saprolite); A_2 B_2 C_2 incision of drainage along fracture zones forms cupola-shaped hills still mantled by regolith; A_3 B_3 C_3 further weathering and erosion reduce hills in size; rock is exposed on C_3 as a result of landslide; A_4 B_4 these hills are continuously reduced by weathering beneath saprolite mantle; C_4 rocky inselberg (dome or bornhardt) emerges from exposure created by landslide.

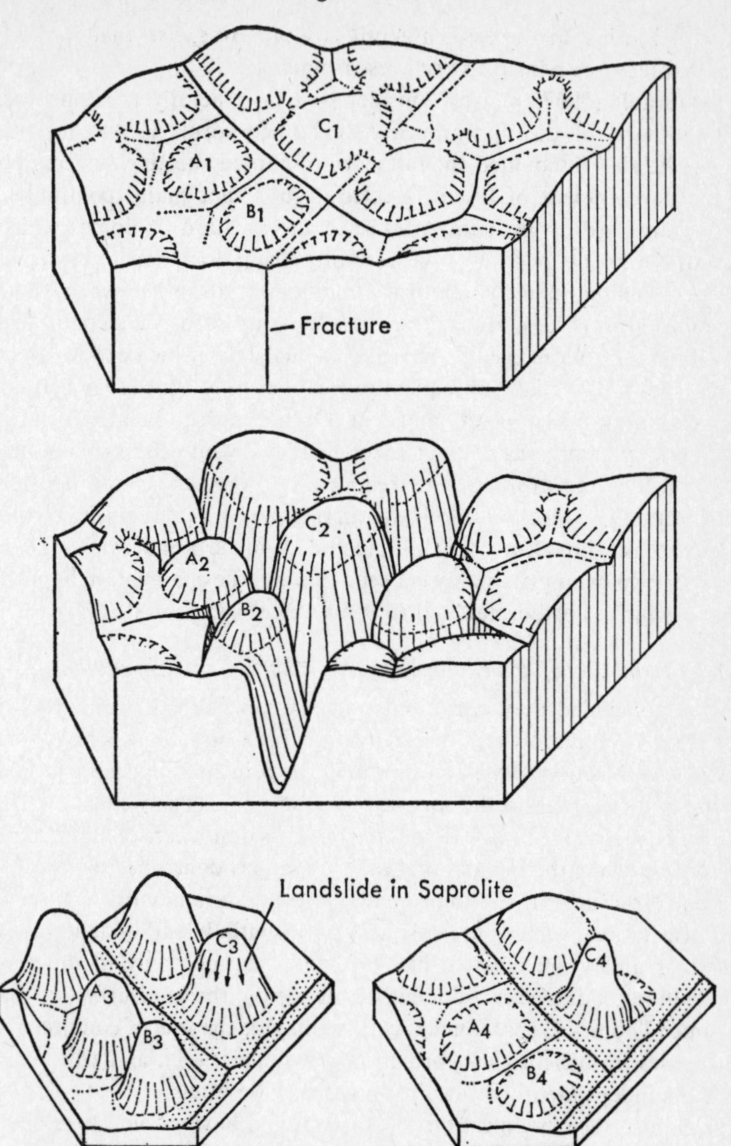

Figure 35 Formation of rocky inselbergs in the forest zone (after Hurault 1967).

For key see opposite page.

exceed in height any known depth of weathering, and that they are zonal features related to cyclic escarpments.

Hurault (1967, p. viii) considered that 'generally speaking this hypothesis of the appearance of exhumed forms is an illusion', and he also discounted the influence of curved sheeting surfaces on the development of inselbergs. He argued that major continuous fractures are of great importance in the location of domes which are 'forms of pure erosion resulting from local and accidental phenomenon, namely removal of regolith'. In other words, like Wahrhaftig (1965), Hurault sees the emergence of bare rock faces as due to random events, particularly landslides which expose areas of bare rock that subsequently weather only slowly and resist re-colonisation by plants and soil. Once exposed, such rock faces gradually emerge as distinct relief features, while the surrounding regolith covered slopes are reduced by weathering and erosion (figure 35). Antecedent deep weathering is not a prerequisite of this hypothesis, but clearly it is not precluded. In the broader view there is difference of emphasis only between this and the exhumation hypothesis which is elaborated and to some extent substantiated below.

The formation of domed inselbergs (bornhardts) by exhumation

It is concluded that, while it is not justifiable to claim universality for the exhumation hypothesis of dome formation, there is no fundamental objection to the application of the principles of deep weathering and stripping to the formation of bornhardts as well as tors. It can be suggested that the domes originate as domical rises in the basal surface of weathering and that they are largely controlled by internal rock structures. If the joint system in an area is likened to a rectilinear lattice, varying in its density both spatially and in depth, then larger and more massive blocks may occur not only in different locations, but also at different depths below the landsurface of the time (Thomas 1965a). Such locations may be the foci of compressive stresses in ancient rock formations (Twidale 1971), and will occur quite independently of local or regional base levels.

Sub-surface modifications of these cuboid blocks could take place by the operation of both weathering and concomitant sheeting, and they would form as nascent domes within the regolith whenever and wherever the rate of groundwater weathering exceeded the pace of surface erosion. The relative relief of these domical rises in the weathering front is unlikely often to exceed 50 m, although along shatter belts or faults alteration may reach to much greater depths.

Although Ollier (1965) has suggested that surviving weathering profiles may be remnants of more extensive and much deeper mantles of the past, there seems to be little evidence to support such an hypothesis. The continuity of weathering processes has been stressed in this account, and their progress is likely to be halted only by great frigidity or aridity of climate, or by rapid diastrophism. Certainly relict profiles occur widely in Australia and elsewhere over the shields, but while they seldom exceed 100 m in depth, it is possible to point to the wide altitudinal range over which deep weathering occurs within a given area. In granite, a tendency for stepped topography to occur is clear from the Sierra Nevada, and from the Snowy Mountains of Australia. In these cases the treads of the steps may be deeply weathered, alternating across a wide altitudinal range with rocky step fronts. This phenomenon led Wahrhaftig (1965) to suggest that exposures would be produced randomly along the mountain front during spurts of uplift. This view substantiates the idea of continuity in weathering, but indicates the oscillatory nature of surface erosion. It also suggests that some domes may arise without the deterministic requirement of structural control.

Exhumation of the sub-surface domes will occur when the relative rates of weathering and surface erosion are reversed. Such conditions may arise, as above, during rapid uplift and incision of streams. When this process gives rise to the formation of steps or escarpments, conditions favouring dome formation may persist as the escarpments evolve in time. Incision of the regolith may equally occur as a result of climatic change, for instance towards more arid conditions with open vegetation cover and more rapid hillslope erosion.

Actual exposure of the domes may arise during stream incision between closely spaced domes (plate 21), and as local mass movements remove the remaining regolith cover (Hurault 1967; figure 35). Alternatively the domes may be revealed during slope retreat within the regolith (Mabbutt 1961a). In the first case, stream incision may extend below the depth of preweathering, so that the initial phase of exhumation is followed by growth due to stream erosion into only slightly weathered rock. In the second case, the initial height of the domes is likely to be limited by the local relief or by the depth of the regolith (Mabbutt 1961a; Thomas 1965b). This process generally leaves the deeper basins or troughs of weathering unaffected (Mabbutt 1961a; Ollier 1965), and these may deepen with continued weathering between the domes which will be increased in height only during a subsequent phase of stripping to lower levels (figure 36). According to this model antecedent weathering occurs before each phase of growth as suggested by Falconer (1911).

Complexities of form may arise during this process in a number of ways. Domical rock masses may emerge contiguous with each other but at different levels, so that, at any time, a range of forms from rock pavements to high domes may occur together. According to many writers periods of lateral denudation may also give rise to rim-pediments around the domes (Büdel 1957).

Summary of evidence for a sub-surface origin of bornhardt domes

(1) The extent, depth and thoroughness of chemical weathering over wide areas of the tropical shields make such a theory tenable.

(2) The detailed form of basal surface of weathering with its discrete basins and domical rises (Thomas 1966a; Enslin 1961; Feininger 1971; figure 20) indicates that the patterns of weathering are favourable to the production of domical and spheroidal forms. This inference can be confirmed from many road and rail cuttings, and there is one recent record of an artificial excavation of a 50 m high, quartz–diorite dome in southern Cameroun (Boye and Fritsch 1973). Analyses of the surrounding materials confirmed that this dome had been embedded in an *in situ* regolith (Boye and Seurin 1973) and the authors suggest that it is located by both petrographic and structural factors.

(3) The commonly observed sharpness of the basal surface of weathering (Rougerie 1955; Birot 1958; Thomas 1965a, 1966a) provides a major break in the erosional mobility of earth materials which is likely to be exploited by agents of denudation, thus leading to stripping of the rock surface.

(4) The fact that some stream channels have been shown to be weathered to considerable depths indicates the feasibility of a theory postulating the 'superimposition' of drainage upon the irregular basal surface of weathering (figure 24).

(5) The apparent correspondence of weathering patterns and joint

systems on the one hand and the very much better established correspondence between dome margins and fracture directions on the other, is circumstantial evidence in favour of joint control over the development of domed inselbergs.

(6) The occurrence of deeply weathered troughs contiguous with domical outcrops, and the steep plunge of the basal surface around many outcrops, together with the persistence of such weathering troughs within widely stripped land surfaces, indicates the spatial association of bornhardt domes with the deep weathering patterns demonstrated before.

(7) The presence of saprolite, and in some cases laterite, blocks on inselberg summits (Falconer 1911; Eden 1971) indicates the former existence of a deeply weathered landsurface of moderate or low relief above the level of the present hills, and from which they could have been exhumed.

(8) Certain sections indicate the presence of corestones within the regolith around individual domes and they may also be seen as surface features resting against the side of the domes. The frequent spatial association of corestones, tors and domes, the merging of the two forms—the spheroidal tor and the domical bornhardt—(Thomas 1965a, 1974 and plates 15, 17) and the presence of tors on dome summits and flanks demonstrates an intimate association of the two forms that is suggestive of common origins.

(9) The domical form would be expected as a result of the modification of cuboid joint blocks by sheeting and chemical weathering (some small domes occur which show few signs of sheeting).

(10) Bornhardts exhibit highly variable morphology such as would be expected from intimate structural control and the operation of many different processes. Features such as vertical faces, a high degree of asymmetry, variable morphology at the junction between hillslope and plain, all argue against a simple application of the theory of parallel retreat to account for bornhardt morphology.

(11) The *individuality* of bornhardt domes can be demonstrated by reference to many granite intrusions, where closely spaced domes are to be found (figure 30, p. 159). This makes any suggestions that the domical form is a result of systematic modification of larger upstanding rock masses difficult to maintain. All the evidence points to domes being formed as domes—not by the shrinkage of larger rock masses.

(12) Finally, the previous observation is supported by the wide variety of topographic sites occupied by the domes: along escarpments, within zones of dissection, interrupting valley sides, as well as on divides. The claimed ubiquity of the outward sloping pediment is quite misleading. In particular, the apparently random disposition with regard to major rivers and divides shows that they are not commonly residuals of landscapes at a certain stage of evolution.

Nevertheless acceptance of this hypothesis without qualification is scarcely justified. Two important possibilities must be kept in mind which, while they do not necessarily contradict the concept of exhumation, do require a shift of emphasis. Increasing knowledge of sheeting or exfoliation as a phenomenon of stress relaxation suggests that, although sheets are not marked on many tropical bornhardts, they may have been formed and the rock plates shed at a remote date. The domical form should not therefore be associated solely with sub-surface weathering; it is probably, as White (1945) predicted, an equilibrium form in massive rocks and may develop from separate causes or from a combination of factors. Also, the acceptance of weathering penetration as a continuous process must allow for the development of domes from chance exposures of bedrock on hillsides (Wahrhaftig 1965; Hurault 1967). The differential rates of weathering as between mantled and exposed rock surfaces could ensure the lowering of the former and the development of the latter into major relief forms (figure 35, p. 193). Although this hypothesis is attractive and allows for the development of domes and other inselbergs from chance exposures following landslides or stream erosion (Wahrhaftig 1965), there is much evidence that contradicts the proposition that all, or even most, bornhardt domes are located by chance events. In fact they appear to be arranged irregularly, and in structurally controlled groups. Nevertheless, structurally located domes could develop in a similar manner, especially in zones or periods of relief development.

THE SUBSEQUENT EVOLUTION OF DOMED INSELBERGS

Landforms developed entirely under surficial conditions are subject to similar processes throughout their histories. But landforms exhumed from beneath a thick regolith emerge as forms exotic to a sub-aerial environment. When buried the basal surface is in reality a weathering front (Mabbutt 1961b) along which fundamental physical and chemical changes are taking place, in the presence of moisture, but in the absence of abrasion. Once stripped of weathered debris such surfaces become virtually arid, for rainfall is subject to rapid runoff and in the tropics residual moisture is quickly evaporated. Under these conditions, little general weathering probably takes place, although some caution concerning this point is necessary as a result of the very high pH values recorded for water standing on such surfaces (Bakker 1960). These records, and the presence of solution hollows on domes show that localised chemical weathering occurs, but it is not such that significant penetration of the rock surface is likely and little disaggregation of rock into gruss is usually found. Thus water flowing over the dome surface is not only dispersed radially, inhibiting channel formation, but is also deprived of an abrasive load.

In this section, therefore, the subsequent development of domed inselbergs will be considered in terms of the separate or combined effects of jointing (including sheeting) and localised weathering. The discussion will demonstrate how these factors commonly lead to degradation of the domed form, although the opinion of Ruxton and Berry (1961b) that the domical form is progressively perfected by continued sheeting should not be entirely discounted.

The effects of jointing

Although the progressive reduction of bornhardt domes has been attributed to sheeting, it has been suggested that this process probably does not continue indefinitely, but ceases once the compressive stress has been released in a system of curvilinear joints. A part of this release appears often to have occurred beneath a deep regolith, but plates are also shed from some sub-aerial surfaces and probably contribute to the perfection of the domical form. However, the weight of evidence supports the hypothesis that the process is self-limiting and is unlikely to lead to major reduction of the dome area or height. Nevertheless, new sheeting systems may be set up parallel to plane joints, when these are exploited by other agencies of denudation.

Under the boundary conditions produced by high relief, incipient or 'potential' joints (Chapman 1958) may open up or become exploited by weathering and erosion. Such joints may transgress sheeting fractures and appear to develop subsequently. The expansive stress operating towards the margins of domes facilitates the collapse of joint blocks from the sides. If this process continues, and works inwards from several directions, the dome must degenerate into a kopje, taking the form of a talus cone surmounted by a surviving monolith or group of castellated blocks. This type of form has been illustrated mainly from Nigeria (Thomas 1965; Pugh 1966; figure 36).

The influence of such joints is not confined to physical collapse: more often they admit moisture and become the loci of sub-aerial weathering. Such joints may generally be divided into lateral and near vertical fractures and lead to the formation of gullies and taffoni (caverns) respectively.

Birot (1958) indicated the localisation of taffoni where sheeting and plane joints intersect, but strong quasi-horizontal weakness appears to be the only requirement. The caverns develop due to the retention of moisture within the joint, although they may be initiated by the effects of runoff. In some cases water issues from within the dome. However once the joint is slightly emphasised, the upper lip on the vertical face becomes shaded and therefore more humid; this in turn induces a faster rate of weathering than on the outer exposed rock face, and therefore leads to growth of the cavern.

Taffoni develop along joints under great confining pressure from the overlying rock, and only occasionally do they appear to penetrate the dome for more than a metre or two. However, they may induce sympathetic sheeting and which leads to some retreat of steep marginal slopes (figure 33, p. 176).

Vertical joints on the other hand lead to the development of weathering hollows which may in turn develop into gullies. There is a considerable literature on these on weathering hollows, but it appears that they may be divided into three basic types (Twidale and Corbin 1963): (1) shallow pans, often irregular in plan but generally broad and shallow; (2) deep narrow pits, semi-circular in plan; and (3) armchair hollows with cirque-like features, open downslope.

The shallow pans do not appear to be related to joints and will not be considered further here. In most cases the deeper pits, however, are aligned along joints, and although Twidale and Corbin did not find all the armchair hollows situated on joints, in my experience the more important ones usually are, and it is these which develop into true gullies. These latter features develop on inclined slopes, and as the near vertical back face is weathered along

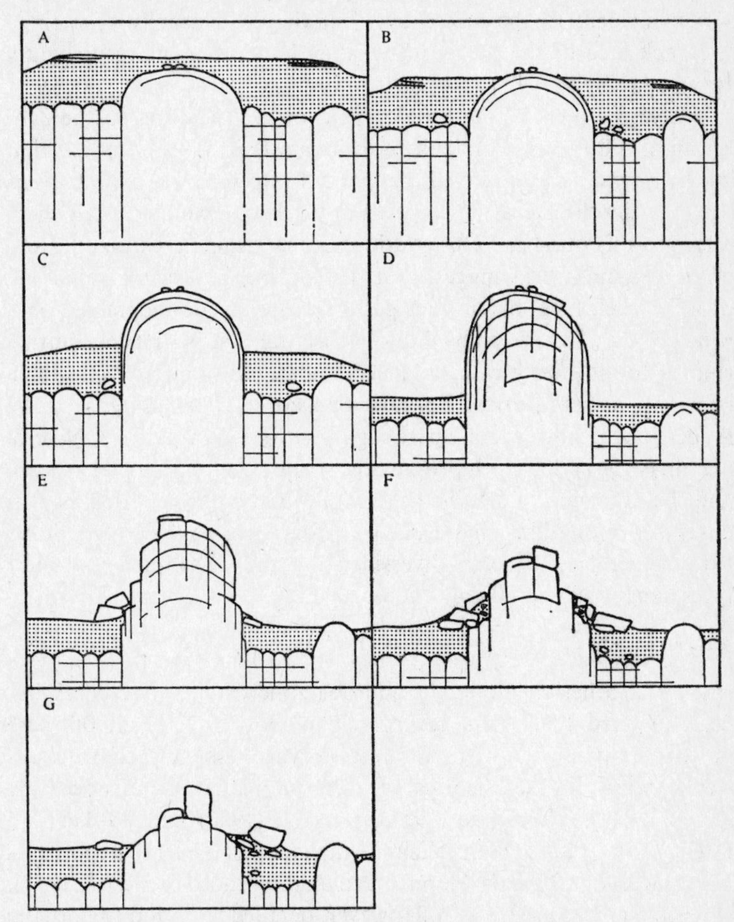

Figure 36 A schematic diagram to show one mode of development and decay for domed inselbergs.

In diagrams A–D both the landsurface and the basal surface of weathering are differentially lowered around the developing domical summit which is stripped of regolith at an early stage. Stages D–G show the disintegration of the dome into a kopje as a result of tensional jointing and peripheral collapse.

the joint plane, it retreats into the dome. The floors of the gullies appear to assume a very low, almost horizontal angle and thus the depth of the features increases with their progress into the hill-mass. They seem to be guided by potential joints that are not always evident to the naked eye, but they can usually be seen in the back face of the gully, where they have been enlarged slightly by weathering.

The importance of the development of such features to the evolution of hill forms is difficult to estimate. Bain (1923) thought that inselbergs might be 'cleaved' or divided into separate culminations by gully development, but few other workers have addressed them selves to this question. Observations along granite escarpments and on domes in Nigeria suggest that the two major processes affecting exposed rock faces are in fact gully and taffoni development, both of which may be assisted by induced sheeting. The gullies concentrate runoff during heavy rain, and accumulate alluvial material which persists between storms. Beneath this cover, weathering is active in deepening and widening the gullies. It was found in Nigeria (Thomas 1967) that both progressive stripping of bare rock surfaces, and the extension of a soil cover within joint controlled depressions continued alongside (figure 32, p. 175). Thus, as with tors, dome surfaces tend to become diversified by the development of new forms under sub-aerial conditions.

Basal sapping processes

Intense weathering along the hillfoot zone around inselbergs has been invoked to explain several morphological features of these hills, and more generally to account for scarp retreat. Such processes were central to the theories of savanna planation advanced by Jessen (1936), Thorbecke (1951), Clayton (1956) and Bakker and Levelt (1964), and are implicit in many other general hypotheses of tropical landform development, including those of Büdel (1957), and King (1948, 1953, 1962). However, detailed attention to both the nature of the hillfoot zone and to the processes operating there has been confined to rather few studies. Most of these have been undertaken in part to explain the sharp 'piedmont angle', found at the base of most inselbergs (not only the domical forms). But they have a wider relevance to the development and maintenance of concave profiles on the lower slopes of inselbergs, and also possibly to the development of marginal depressions or *Bergfüssneiderungen* (Clayton 1956).

Bakker and Levelt, referring to Jessen's ideas, emphasised that 'deep weathering and soil erosion operate with special intensity at the foot of the scarp. Here lies the morphologically most active zone' (1964, p. 28). The contrast between active weathering beneath a moisture retaining soil cover and the very slow changes which take place on the exposed and effectively arid rock face are held responsible for this situation. (Plate 22.) The theoretical arguments of Gerber and Scheidegger (1969), however, suggest that there may also be sound physical reasons for stress release and 'endogenic' weathering at the base of high hillslopes, and this would undoubtedly increase the effectiveness of 'exogenic' weathering by groundwater.

In some cases the abrupt piedmont angle marks a contrast in erosional mobility of the materials comprising the hillslope and the piedmont slope. This may arise simply from the contiguity of rock and regolith surfaces (Ollier 1960; Thomas 1965a); but it may also result from a difference between the size of debris on the hillslope and on the plain. Rahn (1966) suggested for an arid area in Arizona, that the basal angle resulted from lateral erosion by migrating streams, but was maintained by the disaggregation of granite boulders below a threshold diameter. This led to boulder controlled hillslopes having a steep angle of rest and to footslopes of low angle developed across sand and gravel.

In the Sudan, Ruxton and Berry (Ruxton 1958; Ruxton and Berry 1961a) found that the lower, concave hillslopes over granite were commonly waste mantled, and that this regolith became progressively more compact and fine grained downslope (see figure 25, p. 105). Runoff from the upper rock face would rapidly percolate the coarse debris at the upper part of the waste mantle, and a strong lateral flow of groundwater within the regolith would ensue. (Plates 20, 22.) Their conclusion was that 'chemical weathering is most rapid on the hillslope where the subsurface water flows strongest and is most frequently replenished, and most intense on the foot slope where the water lingers longest. It is least active on the bare rock surfaces with no mantling layer of moisture retaining debris and beneath the impermeable clay of the plain' (1961, p. 24). Material from the footslope is removed partly by surface wash but the fine silt and clay by mechanical eluviation, and other constituents in solution (see p. 104). Thus, with intense weathering attack at the foot of the hillslope and effective removal of waste material, retreat of the slope is ensured and a sharp piedmont angle is maintained.

Clayton (1956) considered that the water-table and weathering relationships at the hillfoot zone would bring about basal sapping and scarp retreat, without recourse to arguments concerning the survival of a regolith cover or talus slope in this zone. He cited the occurrence of solution phenomena and incurved faces on bornhardts as evidence for the effectiveness of weathering against bare rock slopes, and argued that a watertable would in fact occur within the unaltered rock, leading to the rotting of the rock beneath the main hillslope. This zone of decomposition would also extend beneath the piedmont slope, and would become subject to selective erosion, following rejuvenation of drainage on the plains. Such rejuvenation would in his view work backwards across the pediment, following zones of more intense rotting, until it reached the hillfoot zone. The extension of the decomposed rock beneath the scarp slope would result in the flushing out of the debris and consequent undermining of the slope, leading to peripheral collapse of overlying rock and to a significant degree of scarp retreat.

Although the field evidence for rejuvenated gullies across weathered pediment slopes is convincing, there has been little evidence to suggest the lateral penetration of weathering far into hillslopes comprised of massive rock. Concave and even 'flared' or overhanging slopes are, however, not uncommon around massive inselbergs. Both Twidale (1962) and Thomas (1965a) have suggested that these mark higher levels of a former regolith cover, beneath which weathering was once active (figure 33).

Slight 'notching' at the piedmont angle in semi-arid areas has been discussed by Mabbutt (1966a), and etched and decomposed rock surfaces are occasionally exposed around the margins of domes (Thomas 1967). But none of these features suggest a mechanism of lateral weathering penetration on the scale implied by Clayton (1956).

On well jointed granite, the horizontal penetration of weathering beneath the hill may be much more effective, and as with the basal sapping and destruction of tors, so also whole hillslopes may be gradually undermined in a comparable manner.

As seen by Jessen (1936) and implied by these observations, basal sapping processes may be of great importance to the study of all hillslope retreat and pediment extension. In the evolution of bornhardt domes it is now clear that several processes may operate; in some cases they will all be effective on different parts of a single hill, in others a particular process may be dominant at certain stages. For instance a dome undergoing outward collapse, due to the opening of plane joints, may subsequently evolve primarily by basal sapping processes. The corollary of these observations is that no general statement concerning the evolution of slopes around inselbergs is of much value: slopes may steepen from basal sapping, collapse and assume stable angles and then retreat parallel to themselves or even decline (Savigear 1960).

The marginal collapse of blocks from bornhardts may be responsible for castellated rock forms generally described as kopjes (or koppies) or sometimes 'castle koppies'. But the widespread occurrence of such forms suggests that not all original originate in this manner. Some may be related to tors and may even possess spheroidally weathered, summit boulders. It appears likely that deep dissection into a well jointed granite may not only expose corestones, but also excavate angular, jointed granite.

Rock pavements

Low rock pavements or 'ruwares' (King 1948; Thomas 1965a) clearly grade upwards into true domes (or large tors), and it may not be possible to define them separately. Nevertheless the existence of such forms remains of interest to this enquiry. King (1948) considered ruwares to be part of the pediment, and therefore a result of lateral hillslope retreat. Büdel (1957) also described them as 'isolated wash pediments' (*isolierte Spülpedimente*), but clearly regarded them as exposed parts of the basal surface of weathering which were maintained at the same level as the plains during slow degradation. However, with rapid erosion, Büdel thought that they would emerge as 'shield inselberge'. In this view they are not necessarily related to hillslope retreat, and in fact they often occur in locations far from major hillslopes.

In general they may be considered as incipient domes or tors, and it is possible that once exposed they are unlikely to be eroded at the same rate as the surrounding weathered surfaces, but that, as in other arguments, exposure of the rock by stripping leads to a normally irreversible process of growth in height and probably area of the exposure. The stripping of almost level rock surfaces is important to subsequent arguments concerning landform development in the tropics, because it indicates that slope angle is not the sole control over the loss of soil and regolith cover. The sharpness of the weathering front is an important factor contributing to this situation which although more common in the highly seasonal savannas is not unknown in the forest climates.

8 Hillslopes and Pediments

Concentration on domed inselbergs as important landforms on tropical shields should not obscure the more general occurrence of hillslopes and piedmont surfaces. These are considered in relation to the question of tropical planation in the next chapter, but, because escarpments and pediments have received so much attention in the literature on tropical geomorphology, it is appropriate to look first at the nature and complexity of major hillslopes on the one hand and of piedmont surfaces, including pediments, on the other.

ESCARPMENTS IN TROPICAL TERRAIN

In many tropical terrains major escarpments are found to separate extensive plains or undulating landsurfaces of greatly differing altitudes. Although King (1953, 1962) and Büdel (1957, 1970) argue from different premises and exemplify different concepts of tropical planation, both have emphasised the importance of erosional escarpments in the relief of the tropical shields. Büdel (1957) sees both the extensive plain and the abrupt, unbroken escarpment as demonstrations of the relative ineffectiveness of stream erosion in the seasonal tropics. King (1953) has also argued that hillslope erosion, particularly in semi-arid areas, keeps pace with gully and headwater incision along escarpments, thus restricting dissection of the upper surface to the depth of incoherent material, or to lines of structural or other weakness. This situation is contrasted by Büdel (1957) with the deep valley development of the temperate and cold lands. Before such theories are considered further it is appropriate to enquire into the prevalence and nature of major hillslopes (escarpments) within the tropics.

In the first place the generation of escarpments as relief features remains an obscure and controversial question. Most of the great escarpments are associated with the southern shield lands which happen also to lie astride or close to the tropics. Yet the Drakensberg extends southwards to 35° latitude, as does the escarpment zone in New South Wales. These features have much in common with the Eastern and Western Ghatts of India (Büdel 1965), the coastal scarps of Brazil and Guyana (McConnell 1968), and the Darling scarp of Western Australia, and all are situated along faulted or monoclinally folded margins of the fragments of Gondwanaland. The tectonically active zones of the tropics (figure 2, p. 5), like their counterparts in temperate latitudes, are dominated by deep riverine dissection. In spite of their impressiveness the southern African scarps are also deeply dissected into spectacular ravines as in the Tugela Basin and the Valley of the Thousand Hills in Natal. Comparisons with the northern shields are made very difficult, because the latter became the centres of Pleistocene glaciation and are still recovering from the isostatic depression that this entailed.

The preservation of escarpments in the interior of the tropical shields has also received comment, but such features are less extensive and less prominent than those of the elevated rims of these landmasses or the great rifts of eastern Africa. Claims that African relief can be likened to a staircase of progressively higher and older surfaces ascending inland from the coasts are a gross oversimplification of a complex situation. However, marked changes of level undoubtedly occur via narrow zones of dissection, or across true escarpments, in many areas of the tropics where active folding has not occurred in recent geological time.

Close examination of many of these escarpments shows, however, that they are closely controlled by geological factors. The Voltaian scarp of Ghana is a good example, for, although the base of this feature appears to be developed in the Basement Complex rocks (Hunter and Hayward 1971), it is nonetheless associated with the occurrence of the resistant Voltaian sandstones. In West Africa, outside of the sedimentary basins most escarpments may be identified with changes in the lithology of the basement rocks. Both the Older (Cambrian) and Younger (Jurassic) granites of Nigeria give rise to some striking plateaux bounded by steep escarpments (figs 30, 31, pp. 169, 171). Scarp retreat into these granite massifs has been minimal. In central Ghana and in Sierra Leone, escarpments around high plateaux are commonly associated with belts of metasedimentary rocks capped by lateritic deposits. Differential erosion is therefore demonstrated by many such occurrences, and as with many residual hills the inference is that the relative relief along the escarpment zones has increased as contiguous surfaces on less resistant rocks have been progressively lowered.

With these qualifications in mind some examination of such escarpments is appropriate, but it is also clear that many of the comments adduced concerning the forms of residual hills and the possible processes responsible for these, are also relevant to the more complex hillslopes comprising major escarpments. Nevertheless it should be realised that many escarpments are complex morphologically, and not many studies have dealt adequately with this complexity.

Clayton (1956) described an erosion scarp in Central Cameroun, some 500–600 m high and extending from east to west for 110 kilometres. It separates the Ndomme Plateau in the north from the Wute Plain in the south and over much of its length is dominated by sheeted half-domes. It was in this area that Thorbecke (1951) described intensive weathering and the formation of depressions (*Bergfüssneiderungen*) in the scarp-foot zone. Clayton (1958) discussed these scarp-foot depressions in detail but gave little information concerning the upper slopes of the escarpment. There is no doubt that such depressions are of great importance in the understanding of the denudation system operating across the escarpment zone, but before discussing them in detail further comments on such escarpments in crystalline rocks are necessary.

Morphological zones across granitic scarps

Study of an escarpment forming the western boundary of the Jos Plateau in Nigeria showed that over much its length, it is divisible into a series of zones, each dominated by a particular process or group of processes. In general, the upper Plateau surface is mantled by laterite, overlying both *in situ* weathered material, and also volcanic and alluvial deposits of Tertiary and Quaternary age. Towards the Plateau edge stripping of these deposits has revealed an inselberg and tor landscape (figure 31, p. 171) within which at the present time both renewed weathering and active stripping are at work. From the Plateau surface at 1160 m this dissected zone descends to 1100 m, below which it falls abruptly as an escarpment comprised of contiguous 'half domes' of bare rock to below 900 metres. At the base of this scarp is to be found a zone of coallescent alluvial fans, ferruginised and indurated in places. Many of the domes near the top of the escarpment appear to be divided by joint planes, into monoliths and kopjes, but towards the base of the scarp they become massive, convex rock spheroids. (Plate 24.) In some areas the fans are obscured or replaced by vast accumulations of boulders, mainly in the form of enormous corestones which have either been lowered gradually with the surrounding plains or may have fallen from the upper zones of the escarpment.

Dissection of this escarpment is confined to a few streams flowing between the domes, but in many cases, as with the Sha River, the rivers flow over the sheeting surfaces of the granite itself, forming no major re-entrant valleys. (Plate 23). Only within the upper, tor-forming zone is there a landscape of ridges and valleys. Outlying hills beyond the escarpment are few. It is clear that lowering of the plains and the growth of the escarpment have been more significant than parallel retreat of the scarp zone in such areas. Many escarpments in Nigeria correspond with this type and are formed where granite stands above basement metamorphic rocks. In cases where the granite is divided by major zones of shattering, penetration of lineal erosion between the joint blocks may divide the mass into closely spaced groups of bornhardts. Comparison of two Older Granite massifs in south-western Nigeria (figure 30, p. 169) suggests that granite plateaux may evolve into groups of irregularly spaced domes by this means.

Nevertheless, slope retreat in granite hillslopes is apparent in some areas. Examination of the footslope area beyond a granite escarpment in northern Nigeria revealed some 200 m of encroachment into the granite outcrop, beyond which convex rocky slopes rise along a sharp piedmont angle. However, it was not possible to identify any contemporary process capable of promoting retreat. In this case a stepped topography, comprised of a series of convex, granite slopes rising from low rock pavements at the foot of the escarpment to form domical culminations capped by tors at the summit, suggests analogy with stepped granite scarps in the Sierra Nevada, described by Wahrhaftig (1965). This author found that the inner parts of the treads were deeply weathered and that the edges of the steps and the risers towards the next level were of exposed rock. On the basis that differential rates of weathering would exist between exposed and buried rock surfaces, the formation, perfection and preservation of the steps would be assured. Wahrhaftig (1965, p. 177) thought that 'such outcrops, formed during spurts of uplift . . . , could grow into escarpments through differential lowering of the country above and below them, and at the same time the area above each escarpment would be flattened to a base level provided by the lowest point on the scarp'. The importance of this analysis lies in the hypothesis that, during periods of rapid erosion granite outcrops could appear randomly and without obvious structural control, as has also been suggested by Hurault (1967). There seems little doubt that many outcrops are structurally controlled, but particularly where open, savanna vegetation permits powerful stripping (plate 25), Wahrhaftig's ideas may have importance. In the case studied, such a general hypothesis can be combined with some degree of jointing control to show that the scarp retreat may be more apparent than real (figure 37).

However, this emphasis on stripping and the operation of processes on bare rock faces becomes less important in more humid regions. Major hillslopes of granite studied in West Cameroun, West Malaysia and New South Wales, appear to be mantled by shallow weathering profiles from footslope to summit. The summit of Gunong Pulai in Johore (710 m) for instance is crowned by two or three small corestones, and the steep slopes leading up to the summit show similar features. Around the footslopes a few minor rock faces appear, and tors are again found within the forested slopes (plate 13). Such slopes may exhibit steep angles of more than 35°. Near Igbajo (figure 30, p. 169) in western Nigeria such corestone slopes are found to the summits of some hills, but in other cases stripping has revealed domes at the summit. Although this is in forest today, the emergence of these dome summits may well be due to a previous period of more arid climate.

Although the jointing characteristics of the rocks will greatly affect the resulting landforms, it may be justifiable to suggest that complete stripping of the basal surface of weathering is less likely within the core of the tropical rain forest climates, where recent climatic changes may have had little effect. Within this denudation system, a thin regolith is able to persist even on steep slopes in excess of 30°. Examples of such conditions can be found in relatively stable areas of Malaysia, and also in the tectonically mobile areas

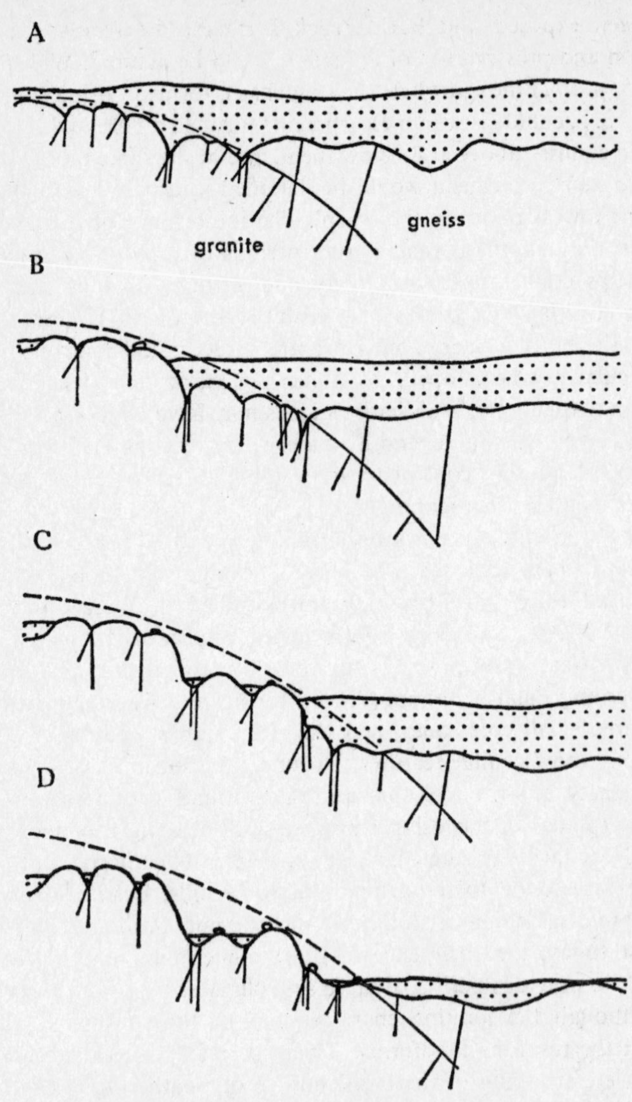

Figure 37 Hypothetical stages leading to the development of a stepped hillslope in granite.

This development is based on the ideas of Wahrhaftig (1965), but is applied here to account for observed hillslope features around the Jos Plateau, Nigeria.

of New Guinea (Ruxton 1967), where linear slopes and narrow knife-edged ridges indicate a uniform rate of erosion across the hillslope, resulting in a parallel retreat of the valley-side slopes and a simultaneous lowering of the ridges.

In the humid tropics the role of footpoint weathering and the development of a piedmont angle are generally regarded as of less significance, because groundwater persists beneath the regolith across the entire hillslope and throughout most of the year. Nevertheless a sharp break between hillslopes and surrounding plains commonly remains a feature in terrains not subject to rapid stream erosion. In Johore, for instance, Eyles (1968) and Swan (1967) have drawn attention to this situation.

Scarpfoot depressions (Bergfüssneiderungen)

It is at this piedmont angle that marked linear depressions appear to develop around many escarpments. These depressions may run more or less parallel with the hillslope or may extend away from it. Not uncommonly, both types are found in a branching system. Although Mabbutt (1965) has illustrated the occurrence of this phenomenon around a single inselberg, Pugh (1956) did not find this common in northern Nigeria. Most observations show that these depressions are developed most clearly in savanna landscapes and around plateaux or other major hillslopes.

Different accounts of these depressions have been offered, and as with many cases of this kind, it seems possible that different kinds of features are involved (Pugh 1966). Pugh (1956) described shallow depressions in the Kilba Hills, Nigeria, which in his view had formed as streams became deflected by developing fans at the base of the hillslope. He related this development to dwindling stream discharge during the progressive denudation of a plateau mass by scarp retreat, but recent short-term changes in discharge, resulting from climatic oscillations, might account for the features described. The depressions described by Clayton (1956) appear to be of a different kind, since they are related to the drainage system of the lower surface, and are not part of any integrated drainage from the upper surface above the scarp. Furthermore, they have the form of gullies, and Clayton argued that they formed by headward extension within a deep regolith, differentially developed at the foot of the hillslope and extending into the plains along lines of geological weakness. Clayton's hypothesis of scarp retreat based on these observations was discussed before (p. 204).

Twidale (1967) invoked a similar hypothesis to explain the sharp piedmont angle and the development of scarp foot depressions on sedimentary formations in the Flinders Ranges of South Australia. He concluded that 'the abrupt change of slope at the piedmont angle is caused principally by differential weathering at the foot of the scarp' (1967, p. 113). In chapter 7 it was doubted whether deep lateral penetration of weathering into the base of the hillslope actually occurs, but there is no doubt that deep weathering of the piedmont slopes is common. The gullies described by Clayton (1956) are readily observed, and weathered compartments in dissected piedmont zones are described by Ruxton and Berry (1961a).

Some processes affecting escarpment zones

It is clear that many individual processes affect escarpments within tropical climates, and their effectiveness in bringing about slope retreat probably depends on many factors, including jointing or other weaknesses in the rocks, the height of the escarpment and the seasonality and humidity of climate, to name but a few. The complexity of scarp morphology is also such that different processes may operate at different levels. For instance across the granite scarps of the Jos Plateau dominant processes appear to operate:

A. *Within the old weathering profiles forming upper slopes* Stream incision, hill wash processes, and continued rock weathering lead to partial stripping, formation and destruction of tor groups, and the development of an irregular tor studded landsurface with occasional large bornhardts (figure 31, p. 171).

B. *Along the upper levels of the rocky escarpment* Removal of regolith keeps pace with or exceeds rate of formation, leading to widespread exposure of bare rock. The upper faces of domes or groups of tors appear to undergo outward collapse, leaving a castellated profile of divided domes.

C. *On the mid-slopes of the scarp face* Massive half domes appear to undergo sheeting as a result of stress relaxation, shedding plates to lower levels. Steep gully heads extend between the dome faces, but seldom dissect the scarp into deep valleys or embayments.

D. *On the lower slopes of the escarpment* Great accumulations of talus and/or corestones suggest both decay of the upper slopes and accumulation of cores from marginal weathering of the escarpment at different stages in its vertical development. Elsewhere, fan accumulations mark the effects of former debris-laden streams, or dome faces may intersect the surface of the plains at a sharp piedmont angle.

E. *Within the weathering profiles of the footslopes* Intense weathering and selective stream erosion commonly lead to weathered plains as described above, with possibilities of basal sapping. But in some locations accelerated stripping of the basal slopes has exposed rock surfaces. (Plates 25, 26.) Under these conditions the base of the escarpment is attacked by gully heads following joints in the rock and by sympathetic sheeting.

Since these remarks are entirely qualitative, although based on field observation and mapping, it is not possible to construe them to support or refute hypotheses of scarp retreat or decay. Evidence for the former must come from comparisons of erosional and petrographic boundaries. These suggest that in some cases retreat is more apparent than real, but in others it is quite unambiguous. These situations are usually found in layered sedimentary rocks or in well jointed granites.

PEDIMENTS AND SIMILAR PIEDMONT SLOPES

The pediment landform has occasioned a great deal of comment and much controversy ever since the term was introduced into the literature of geomorphology by McGee in 1897. Like the French term *glacis*, it refers to a plane or gently concave surface, ideally undissected by gullies or other permanent stream channels. Both terms derive from architectural usage* and their descriptive meaning in geomorphology has always been ambiguous. Many writers insist that pediments occur only in arid and semi-arid climates, or in areas which have formerly experienced such conditions (Tator 1952, 1953, Mackin 1970, Twidale 1968, etc.). But following the concepts of

* *Glacis* were gentle slopes leading from the walls of seventeenth century towns to protect them against cannonballs; pediments were gently inclined gables surmounting the porticos of Greek temples.

W. Penck (1924), authors such as King (1953) have argued that pediments result from slope retreat, and are present in all climates. Others, such as Pugh (1956, 1966) have claimed that pediments comprise most of the landscape within the savannas. In part these contradictions result from varying definitions of the term.

Pediment definitions and characteristics

Birot (1968, p. 96) comments that pediments 'have properties that are intermediate between those of a slope and a river bed', and this does focus attention upon the traditional use of the term to describe surfaces of transportation over which the principal process is that of overland flow.

However, definitions of pediments, like those for laterite, vary greatly. The term may be restricted to the 'piedmont surface that cuts across the rock formations of the mountain ranges' (Tuan 1959, p. 3), or may be widened to include any 'relatively unrestricted degradational surface produced by subaerial agencies' (Tator 1953 p. 49). There is a tendency amongst many writers to equate pediments with rock cut surfaces carrying only a superficial cover of alluvium or regolith. The presence of a sharp 'piedmont angle' at the back of the pediment is often considered necessary to pediment definition except where coalescent pediments adjoin. This break of slope is not always as evident in humid tropical regions as in semi-arid areas. But nonetheless it is commonly recognisable, and in Johore, for instance, it is very marked (Swan 1970a). Though Lester King has encouraged the association of pediment forms with lateral rock planation his own definition of the term serves as a basis for many different observations and interpretations. King (1962, p. 143) described the pediment as 'a broad concave ramp extending from the base of the other slope elements down to the bank or alluvial plain of an adjacent stream. Frequently its profile approximates to a hydraulic curve and it is unquestionably under the action of running water. Beneath the surface wash the bedrock is usually sharply truncated by erosion, but sometimes the subadjacent rock may pass upward through progressively weathered mantle into soil. The concave surface profile has then probably developed only in previously weathered materials. Of course landforms that were originally cutrock pediments may, if the cycle of hillslope development is arrested, deteriorate into the second condition as the transported mantle is slowly removed and weathering slowly proceeds beneath it into bedrock.'

Such a definition may guide the present discussion, but should not carry the implication that all such surfaces are essentially cut rock features, or of necessity fashioned by overland flow. Pallister (1956) encountered this problem in describing pediments in Uganda, which were developed across the lower horizons of a deep, lateritic weathering profile. He retained the term 'in view of the essentially genetic relationship believed to exist between these slope elements and the more easily recognisable topographic features of arid climates' (p. 82, footnote).

This usage of the term raises some problems of terminology, because continental, especially French, geomorphologists have used the term *glacis* to describe erosional slopes in unconsolidated materials. Birot and Dresch (1966) refer to three major types of glacis:

(1) glacis of erosion which truncates soft rock outcrops and has only a thin veneer of transported debris;

(2) buried glacis, where old erosional surfaces are overlain by later alluvial or colluvial sheets of transported material;

(3) glacis of pure accumulation, such as fan surfaces.

Furthermore many authors require pediments to occur as 'piedmont plains' (Mensching 1970) and therefore to extend from the base of major hillslopes. In the case of pediments developed across the weathering profile, these commonly extend only from the foot of low breakaways. Mackin's (1970) requirements for defining pediments seem the most satisfactory:

(1) they are surfaces of erosion developed across either hard or soft rock;

(2) they should be planar or of low relief;

(3) their overall slope should lie in the range of tens or hundreds of metres per kilometre (usually between 1° and 7°);

(4) their slopes should decrease uniformly from the mountain mass (if present) towards local base level.

Mackin (1970) regretted the application of the term to all piedmont surfaces, and considered the landform to occur only in the arid and semi-arid zone, and in the absence of mass movement. However, the term is applied also to savanna and even rain forest areas of the tropics (Pugh 1956, 1966; Swan 1970a), where at least some of the definitive characteristics listed by Mackin appear to obtain. But the role of running water under a dense vegetation cover must be questioned, particularly on slopes as gentle as 1°–7°.

The justification for the use of the term in more humid regions rests not so much on any assumed identity with arid regions in terms of process, as on a general similarity of form and a particular relationship with contiguous slope facets in the landscape. Broad, gently sloping surfaces which extend from the base of hillslopes, and which in some cases pass beneath alluvial accumulations and in others terminate at a break or slope leading down to the river channel or floodplain, undeniably exist within the tropics. King (1953, 1962) has maintained that all these are replacement slopes, extending as the hillslopes retreat. But whatever their genetic relationships with hillslopes or connections with planation processes, these slopes form surfaces of transportation. Where they are developed across incoherent material in humid areas, it has been suggested that they are relict forms from a former semi-arid climatic phase (Rhodenberg 1969, 1970). This arises partly from the view expressed by King (1953) that pediments are equilibrium forms where intense tropical rainfall must be dispersed rapidly across poorly vegetated surfaces. Although undissected, pediment-like surfaces occur in the humid tropics, piedmont slopes are more commonly dissected into undulating lowland, and as shown in chapter 6, this type of terrain is very widespread in the forests and wetter savannas. Furthermore, although overland flow has been shown to occur on steep slopes under forest, it is probably negligible on gentle slopes of piedmonts.

Tator (1952, p. 123) makes an important observation in this context: that 'the pediment containing regime must of necessity be such that lateral erosion and weathering is promoted, but extensive vertical activity in this sense is prevented'. An absence of strong vertical erosion by streams is clearly axiomatic if pediments are to be extensively preserved and actively extended, but the role of weathering in this context may be fundamental.

Many writers (Handley 1952; Pallister 1956; Mabbutt 1961a; Thomas 1965a; Eden 1971) have noted an apparent absence, over wide areas of the tropical shields, of rocky pediments. A great deal of work in both Africa and Australia has emphasised the truncation of relic weathering profiles by current processes and slope forms.

Thus, many pediments are cut across preweathered materials, and this is a different situation from that which would arise from the deep weathering of truncated rock surfaces, resulting from slope retreat. Many apparently rock-cut pediments are little more than narrow benches around contemporary hillslopes (figure 20, p. 92). Mabbutt made the point, in his study of a part of the Western Australia shield, that 'pediments generally belong to a late stage in exhumation' (1961, p. 111). In a similar study from central Australia he concluded that 'what has been involved in pedimentation has been slight back-trimming at the hill base, partial levelling of an irregular exhumed weathering front, and the imposition of the continuous concave profiles of sub-alluvial and part-subaerial rock surfaces' (1966, p. 91). Büdel (1957) refers to a similar situation, where 'rim pediments' are seen to border highland masses which lead down to deeply weathered plains (figure 40, p. 232). Similarly, I myself (Thomas 1966a) concluded from a study of weathering profiles and jointed inselberg groups on the Jos Plateau of northern Nigeria, that the patterns of deep and shallow weathering in relation to outcrops indicated that the weathering pattern was the primary feature of the terrain, and that the lateral planation of residual hills constituted a subsequent or secondary process following a certain degree of incision, and stripping of the regolith.

On the other hand, there are some landscapes where exposed or near exposed rock surfaces abound, and these can often be explained more adequately by reference to extensive stripping of weathered material than by slope retreat. This assertion appears justified because the slope relationships in certain areas are not compatible with pedimentation, and also because the residual hills show marked structural control over their outlines. It must be recalled, however, that intense weathering may occur at the scarp foot zone, and that, particularly in semi-arid areas, weathering depth may *decrease* away from the escarpment (Twidale 1967). This phenomenon has also been noted in more humid areas (Clayton 1956) and records from southern Nigeria confirm deep weathering beneath talus slopes in jointed granite (figure 38).

These arguments apply most obviously to areas of crystalline rocks, where a distinct weathering front is observed. However, the truncation of deep weathering profiles over sedimentary rocks by the extension of pediment slopes is equally well documented, for instance from Australia (Mabbutt 1965; Hays 1967; Twidale 1967). The discussion of the pediment problem, as with so many other controversial topics in geomorphology, requires enquiry into two sets of facts: those which concern currently operating processes, and those pertaining to the inheritance of landform features, such as deep weathering profiles, from past periods. Mabbutt's study of pediments in central Australia (1966) demonstrates this principal clearly, and the study of a bornhardt dome in southern Nigeria by myself (Thomas 1967; figure 32) further emphasises this difficulty.

Although King (1962) indicated that parallel slope retreat would operate most efficiently with respect to massive, crystalline rock faces there is little evidence to support this assumption. On the contrary, it is the existence of these massive bornhardts that has led to hypotheses based upon the periodic lowering of the weathering front and the land surface around them, leaving the residual domes almost untouched by lateral planation. On the other hand, hillslopes in rocks which retain a superficial cover of regolith and soil, or which are broken into joint blocks of small dimension, undoubtedly undergo forms of retreat which bring about the development of a replacement slope, or pediment. The role of basal sapping or footpoint weathering in this development has already been emphasised, and the action of surface water flow in humid areas doubted. Nevertheless many studies emphasise the role of the pediment as a slope of transportation and this is thought to account for the low angle of slope characteristic of these features. By contrast, the adjacent hillslopes can often be described in terms of the angle of repose of angular debris. Many of the well jointed granite hills of the Jos Plateau appear to have their piedmont junction angles controlled in this manner.

Detailed examinations of pediment profiles have come mostly from arid or semi-arid regions (Ruxton 1958; Mabbutt 1966a; Tuan 1959; Cooke 1970).

Tuan's studies (1959) made in the Arizona desert, concentrated upon the interpretation of forms in terms of contemporary processes. In most cases the granite pediments showed marked relief, and many features associated elsewhere with sub-surface, groundwater weathering. Around the Tortilla and Black Mountains he recorded a granite 'pediment' with an internal relief of as much as 30 m, within which piles of granite boulders commonly reached heights of 6–10 metres.

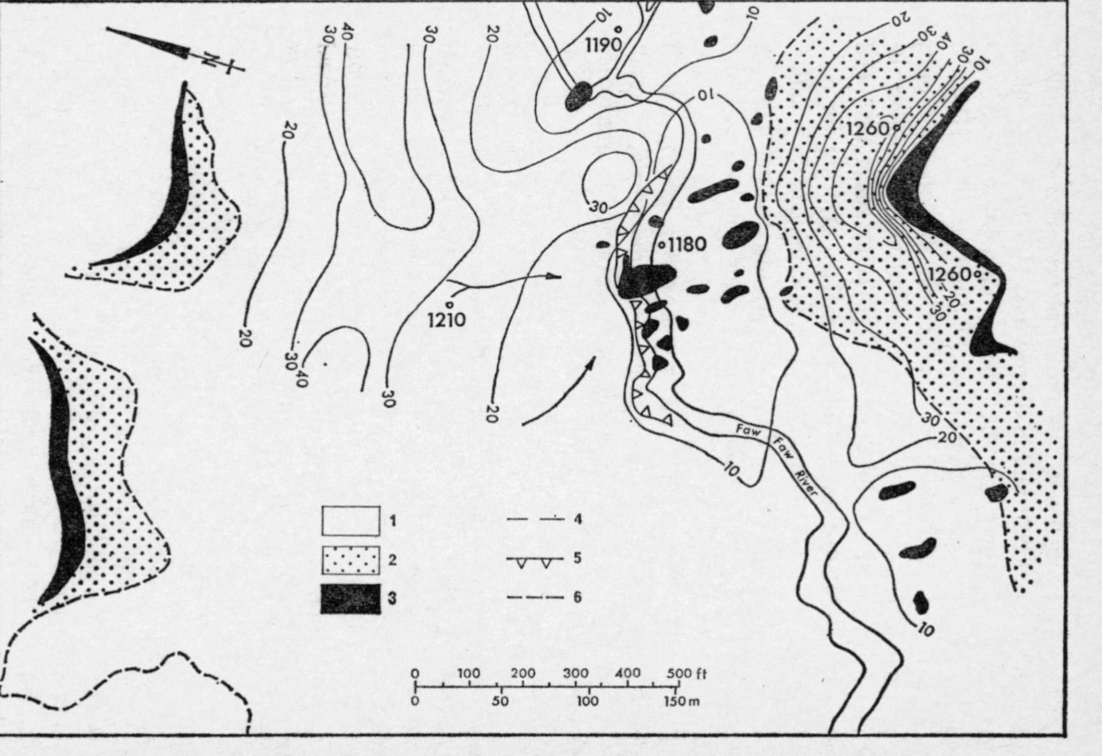

Figure 38 Sub-talus weathering in Precambrian granite near Shaki, Nigeria.

1. Soil covered surfaces; 2. Talus and other boulder-controlled slopes; 3. Outcrops of fresh rock; 4. Contours at 10 feet intervals; 5. Convex break of slope; 6. Concave break of slope around the base of the talus slope.

These tors were found not only on interfluves but particularly in structurally aligned belts and in depressions. This 'tor and boulder' topography was repeated in the Catalina Mountains where Tuan found that the piedmont angle was not clearly defined. Here the relief decreased outwards from a hilly core, but the landforms remained unchanged. This situation recalls that found around the dome examined by me in southern Nigeria (Thomas 1967 and figure 32, p. 174). However, Tuan adduced no evidence for deep weathering in the granites, and thought the rounding of the boulders was subaerial. The relief of the pediments was accounted for by the stripping of a former alluvial cover, some remnants of which were found on isolated hills.

From these accounts and the accompanying photographs it appears possible that the granite pediments in this area are a result of differential etching of the granite, whether or not a former deep regolith is implied. In the Sierrita mountains where the transition from the granite pediment (2½–4°) to the hillslope (28–30°) is most abrupt, Tuan found that a textural difference in the granite coincided with this transition. In this case, as in so many others, evidence for lateral planation of bedrock is ambiguous, and periods of lowering partly through stripping processes may be implied, with the probability of former sub-humid climatic conditions having been influential in this area.

Comparison of this study with that by Mabbutt (1965) in central Australia is instructive. He found an irregular piedmont surface formed over both granite and schist, and considered that lateral stream planation could not account for this phenomenon. In his opinion streams would not be competent to erode the bedrock surface, but would on the other hand lead to incision of weathered materials. He found evidence for 'alternating mantling and stripping with shallow erosion, and of partial notching of bedrock surfaces, commonly at more than one level' (1966, p. 78). Like Tuan (1959), he found that the mantle was colluvial or alluvial and not residual, but he concluded that the piedmont angle was emphasised mainly by sub-soil notching through weathering processes. Some residual weathered material from the granite was found to be poor in clay and Mabbutt suggested that eluviation of clay in the manner outlined by Ruxton (1958) may be responsible. He also considered the role of past processes, and suggested that exhumation of the hillslope from a former weathering profile was responsible for the gross form, and that subsequent events had led to only a limited amount of slope retreat.

In a recent study of pediments in the Mojave Desert, Cooke (1970) raises a number of interesting questions concerning pediment forms and their relationships with adjacent hillslopes. He emphasised the need to consider the entire erosional–depositional system of the pediment, including the mountain area above and the alluvial plain beyond the pediment itself. A number of surprising results emerged from morphometric analyses of these elements. The mean area of pediments in this region is only 5·95 square kilometres and they occupy only 9·4 per cent of the plains area. Their slope relationships are comparable with pediments elsewhere, ranging from less than 1° to around 6°, whilst the hillslopes in quartz monzonite have a mean slope of 28° 40′, but vary from 12° to 31°. Cooke found no significant correlation between proportions of mountain and pediment areas in the Mojave Desert, and tentatively concluded that 'as the pediment becomes larger it does not necessarily do so at the expense of the associated mountain. For example the pediment might become larger at the expense of the sub-alluvial bench' (1970, p. 36).

It is clear from a comparison of these accounts that simple statements concerning the systematic extension of pediments as a result of a continuing process at the present level of denudation may be misleading, even within the semi-arid zone where weathering processes are to some extent inhibited by lack of water. Similarly the assumption that vast areas are comprised of truncated rock surfaces is not warranted: much of the surface appears to be depositional and conforms more to the concept of the glacis than to common definitions of pediments.

Piedmont slopes in humid climates

Pediment studies within the seasonal or humid tropics with which the present study is concerned have been remarkably few. But the recent study by Swan (1970a) of piedmont slopes in Johore is of considerable interest in this context, for although the basal knick or piedmont angle is not as clear as in the arid zone, a similar sequence of hillslope, piedmont slope (pediment), and clay plain is recorded. The pediment slopes range from 1–9°, above which hillslopes range widely from 10° to more than 40°. The pediments display evidence of weathered bedrock at a shallow depth, and Swan concluded that they have extended as a result of backwearing of the steep hillslopes, and that they carry a colluvial or an alluvial cover of material removed from the hillslope.

Evidence from other tropical areas suggests that the hillfoot zone is commonly one of deep weathering, and dissection. Ruxton and Berry (1961a) make specific reference to this situation in an important paper dealing with the contrasts in granite relief forms between arid and humid tropical regions (figure 39). Whilst in the former, the footslopes or pediments below major escarpments display a low relief over shallow weathering profiles, and the removal of clay from upper horizons to form clay plains, the humid areas commonly

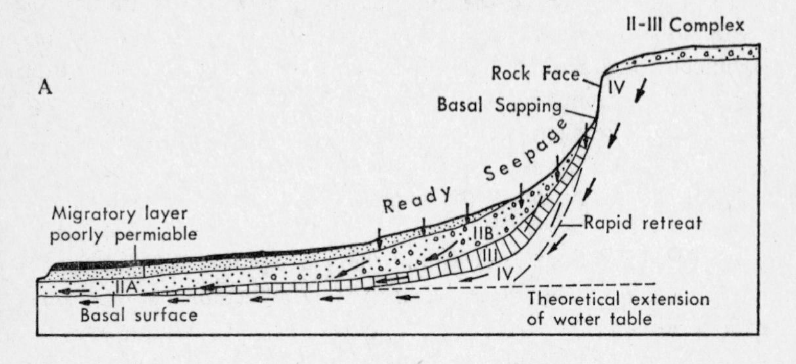

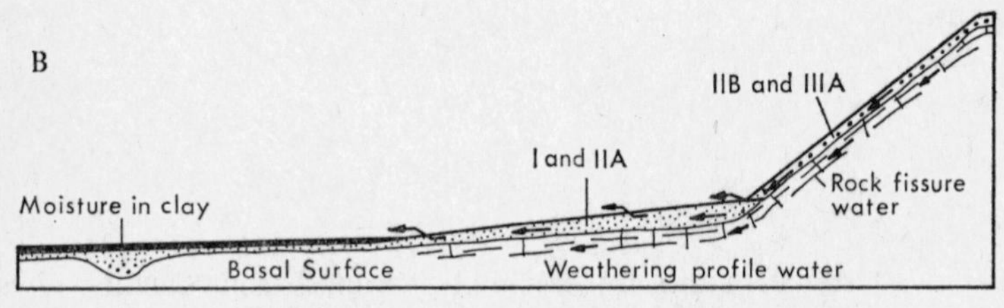

Figure 39 Weathering processes and the maintenance of the piedmont angle (after Ruxton and Berry 1961a).

Showing sub-surface water regime on granite in:

A. Humid tropical regions (based on Hong Kong); B. Semi-arid tropical regions (based on Sudan – see figure 25).
Water flow is indicated by arrows.

exhibit thick lateritic weathering profiles, dissected to depths of up to 150 metres. This dissection produces a distinctive relief, characterised by 'convex compartmented hills with sub-accordant summit levels and delimited by stream and valley patterns following structural frameworks result(ing) from the dissection of the thick weathering profiles which mantle the valley side strips' (1961, p. 21). The evidence for this comes mainly from Hong Kong, and the situation is clearly different from that described by Swan (1970a). This difference may reflect contrasts in tectonic history between the field areas, and in particular the absence of available relief for the development of a deeply dissected lower storey landscape in southern Johore.

This study by Ruxton and Berry, however, does focus attention on the pediment problem within the seasonal tropics, and one of their conclusions echoes comments made earlier concerning some of the classic areas for pediment formation: 'almost every part of every pediment that we have seen or that has been figured in the literature on granitic rocks shows the lower weathering zones not the basal surface (continuously solid rock) at the surface' (1961, p. 22).

Büdel (1970) has recently suggested that in arid areas we may think in terms of *active* slope retreat with the *passive* extension of plains in the form of pediments and glacis carrying allochthonous rock waste from the steeper slopes. In humid areas on the other hand the surface of the plains is *actively* formed by weathering processes, tending to undercut the hillslope which retreats *passively*. Whilst this ignores the important role of weathering on the hillslope itself, beneath a shallow mantle of coarse debris (Ruxton and Berry 1961a) it does suggest that our view of the pediment slope should be that it is a particular type of transportation surface developed under conditions of open vegetation cover and strong sheet wash (King 1953), and that it merges into depositional glacis on the one hand, and into dissected and weathered piedmont plains or lower storey landscapes on the other. There can be no unique association of pediment with retreating hillslope, and it is a mistake to regard all piedmont plains as pediments (Mackin 1967).

This situation is further complicated by the results of dissection and slope retreat *within* the deep weathering profiles. One reason for this is that, as a result of the differences in erosional mobility between regolith and crystalline rocks, incision of the landsurface may produce *two* replacement slopes. Mabbutt (1961a) found that stripping of regolith was controlled by the depth of the weathering profile, and there can be little doubt that retreat of slopes within unconsolidated materials, especially below a cap rock such as laterite, will proceed more rapidly than planation of a rocky relief. Thus, from a single major period of drainage incision two kinds of surface may develop. This was described by me (Thomas 1965b) and more recently by Rhodenberg (1970), who distinguished between 'slopen 'slope pedimentation' within regolith and non-consolidated rocks and 'valley bottom pedimentation' extending from the floors of large valleys. According to this author the distinction is the more important because, slope pedimentation as a contemporary process is confined to semi-desert environments, and thus, where such landforms are found within the savannas of West Africa they indicate climatic change. The rivers of the wetter savannas and forests are now able to dissect these surfaces, and according to Rhodenberg this dissection has continued since the last arid phase of the Würm (at least 30 000 B.P.*).

In concluding this section, therefore, it is perhaps appropriate to suggest that, whilst there is much evidence for hillslope retreat in humid as well as in semi-arid climates, the nature of the replacement surfaces which result from this retreat vary considerably, according to lithology, climate and vegetation, and according to their altitude above regional and local base-levels. Not all such surfaces are recognisable as pediments according to usual definitions. Futhermore, many piedmont surfaces are not so much replacement slopes due to scarp retreat, as surfaces resulting from the lowering of the plains by differential weathering and erosion. Surfaces which we choose to call pediments or glacis should be recognisable as such, and the dissected lower storey landscapes of the humid tropics scarcely fall within this category. Neither do many of the multi-convex rocky surfaces which result from accelerated stripping of regolith. On the other hand it is restrictive to insist upon overland waterflow as a definitive process in pediment formation. As with the domed inselberg, a particular form, in this case the gently concave profile, may not after all reflect a unique combination of processes, but may develop in response to a variety of controlling circumstances.

* All dates are given in years before the present century (B.P.).

It is also clear that climatic change has been important as an influence over the morphology of many semi-arid areas, as well as in the savannas and rain forest. Thus, in the more arid areas relict weathering profiles have apparently undergone advanced stripping, leaving rocky surfaces to be modified by sub-aerial processes, whilst in the humid regions the pediment surfaces developed across old regoliths are currently becoming dissected and modified by mass movement to produce multi-convex landsurfaces.

9 Tropical Planation

Intrigued by the apparent effectiveness of planation in the tropical zone, geomorphologists have returned again and again to the problems of interpretation which these landscapes pose. Already it can be seen that many of the problems are connected with the hypotheses advanced to account for the origins of individual hills. Büdel, whose earlier work has been influential in the study of tropical landscapes, has recently returned to this problem (Büdel 1970) and has emphasised that 'too little notice has been taken of the fact that surface and scarp form an erosion system belonging together' (p. 1). Many writers have focussed upon footpoint weathering which forms only a part of this system. Rather few studies actually examine the relationship between hillslope and plain, and among these the work of Ruxton (1958) remains perhaps the most important. The recent work on pediments by Cooke (1970) also stresses the importance of pediment : catchment relationships. But in order to advance this type of understanding further, a greater knowledge of hillslope hydrology is required. If the nature of water flow across hillslopes and over and beneath the piedmont surfaces was fully understood then a systems approach to the denudation system would become possible. Most hydrological studies, however, have been concerned with either soils *per se* or channel flow, and unfortunately the study of fluvial systems has so far offered only a tenuous connection with the landform as such. Thus, in reviewing the problems of relief development in the tropical zone, the approach must still be largely morphological, and the study of process a guide to an understanding of morphological changes.

It is appropriate at the outset to make certain important distinctions. In landscapes which display two (or more) storeys (Ruxton and Berry 1961a; Swan 1970b), it is implicit that major hillslopes adjoin lower landsurfaces along a narrow zone of abrupt transition. In Africa and in India (Büdel 1965, figure 29) such zones of dissection may separate upland and lowland plains of similar character, but in Malaysia the upper storey of hills and mountainous terrain preserves only a few remnants of upland surfaces, and the lower plains are partially depositional in character, although denudational surfaces are also present. Patterns of erosion and deposition vary not only with climate but also with tectonic mobility and relief development. Thus, on old lands, widespread slope deposits mantle the denudational surfaces, but in young, mountainous terrain the spatial separation of erosional slopes in the mountains and depositional surfaces comprising the piedmont zone is most striking. The landscapes of Johore discussed above and by Swan (1967, 1970b) would appear to occupy an intermediate position within a series of erosional landsurfaces, extending from those almost entirely mantled by superficial accumulations (Ruhe 1956; Mulcahy 1960) to those in which erosional slopes may be free of superficial accumulations of wide extent or great thickness.

These observations are important, because the style of subsequent relief development will be greatly affected by such situations. It is of course difficult to write of 'initial forms' in old landscapes, but it is necessary to consider the events which may have led to the generation of escarpments for these provide clues to many of our problems of interpretation. It seems necessary that the height relationships (really differences) between depths of weathering, valley incision, and escarpments should be regarded as fundamental to any consideration of landform evolution (Thomas 1965b).

It is clear from the literature and from field evidence that discontinuities in the landscape as between hillslope and plain, between rock and regolith, regolith and transported sediment, are of fundamental importance to an understanding of relief development and spatial organisation of landforms. Discontinuities of lesser character as between zones within a regolith, or between rocks of varying resistance are similarly important, for they commonly mark breaks of erosional mobility that lead to differential erosion on varying scales: from breaks and inflections of slope on a hillside to the juxtaposition of plains and hills.

Although, as early as 1896, Branner attested to the importance of deep weathering in Brazil, Falconer is perhaps the first writer in English to advance the concept of widely and deeply weathered landsurfaces in the tropics and to emphasise the irregularity of the weathering front with its rocky projections above the weathered landsurface (Falconer 1911, see p. 190). In the 1930s, however, several writers returned to this theme, amongst the most important of whom were Credner (1931), Wayland (1934), and Jessen (1936). At this time a number of German writers became interested in the tropics and work by Passage (1928) and Sapper (1935) underlies many of the hypotheses of more recent writers such as Büdel (1957) and Louis (1964).

Wayland's concept of the 'etched plain'

The work of Wayland (1934) is particularly important to the understanding of shield landscapes. Working in Uganda at a time when the ideas of W. M. Davis were prevalent he could not reconcile the coexistence of a succession of well developed 'peneplains'. Furthermore, the relationships between these plains suggested to Wayland that they had developed out of each other, without any intervening period of major uplift and formation of diversified topography. Wayland (1934, note 376) commented on this: 'Absence of any marked relief, a flat gradient and a seasonal climate lead to vertical rather than horizontal movements of ground waters and the consequent rotting of all but chemically resistant rocks such as certain quartzites to depths of tens of feet. This zone of rotted rock or saprolite is largely removed by denudation if and when land elevation supervenes, and the process may be repeated again and again as the country rises slowly and continuously. Thus, the surface (or large parts of it whose areas are determined by drainage basins) is lowered and kept at or near base level by superficial removal against elevation below.' Such plains were regarded as 'etched plains, and as such are indicative not of tectonic stability and quiescence but of instability and upward movement'. Altitudinal differences between such plains were thought to be due to 'relatively rapid vertical movements punctuating the slow discontinuous rise'.

Attention was also drawn to the separation of the weathering surface from the erosional landsurface by Credner (1931), but it is to Büdel (1957) that we owe the clearest exposition of the concept of double surfaces of levelling (*Doppelten Einebnungsflächen*). Realisation that over many crystalline rocks two, three dimensional surfaces of fundamental erosional importance exist, gives added point to the recognition in the landscape of slope forms developed over rock and regolith.

Büdel's concept of 'Doppelten Einebnungsflächen'

In the seasonal tropics, according to Büdel (1957), denudation begins with deep decomposition and is followed or accompanied by slope wash processes. The enfeebled streams achieve little valley incision and plain formation results. Such plainlands terminate abruptly against escarpments and around isolated inselbergs with steep, rocky walls. The erosional scarps are characterised by embayments where the plains penetrate slowly by slope retreat, while the rivers draining from the upper surfaces are characterised by rapids and waterfalls.

Current plain formation is taking place today close to sea level, under the influence of sheetwash thought to be particularly active in the seasonally humid tropics, described by Büdel as the *Flächenbildenzone*. This zone has experienced the long continued influence of tropical conditions, which have promoted deep decomposition of rocks. This particular point is of some importance, and Douglas (1969) has emphasised that the inner zone of the humid tropics (perhaps 2 or 3° N and S of the equator) may have experienced persistent humid tropical conditions for several million years, and may be one of the few areas where climatic change has not been influential in landform development. Although this is possibly true, the latitudinal extent of such a zone must be small, and throughout the greater part of the tropics evidence for oscillations between more humid and less humid climatic conditions are well established.

Büdel considered that tropical plains form by extension from lowland coasts or from interior basins and under the conditions already elaborated. He also recognised that plain formation in extra-tropical areas may have occurred under a tropical climate, a point explored in greater detail by Bakker and Levelt (1964) and by Gellert (1970).

In this study Büdel (1957) propounded his influential concept of double surfaces of levelling (*Doppleton Einebnungsflächen*). The separation of the upper wash surface (*Spül-Oberfläche*) from the basal surface of weathering (*Verwitterungs-Basisfläche*), by at least 30 m of decomposed rock, creates a major contrast with temperate areas, where weathering and erosion are sometimes considered to proceed *pari passu*. Tropical plains exhibiting such double surfaces of levelling form gently concave lowlands between enclosing escarpments, and are occasionally interrupted by inselbergs (figure 40). On the upper surface, transportation of weathered debris is carried out by streams and surface wash, but the fundamental denudation is achieved by the weathering of the basal surface into a highly irregular relief which can be exposed by surface stripping. Büdel emphasised repeatedly the identity between plain and river, with their comparable lateral and longitudinal gradients. But close to escarpments these low gradients rise to 3 per cent over rock cut pediments developed by powerful sheetwash from higher slopes. Intermittent lowering of the upper wash surface brings about a series of such rim wash pediments (*Rand Spülpedimente*), notched into the mountain front to

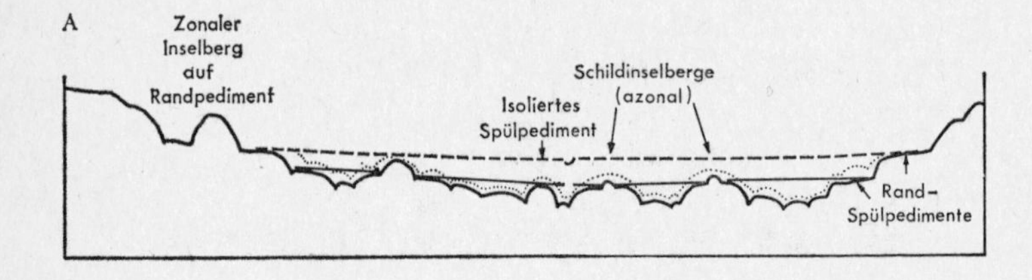

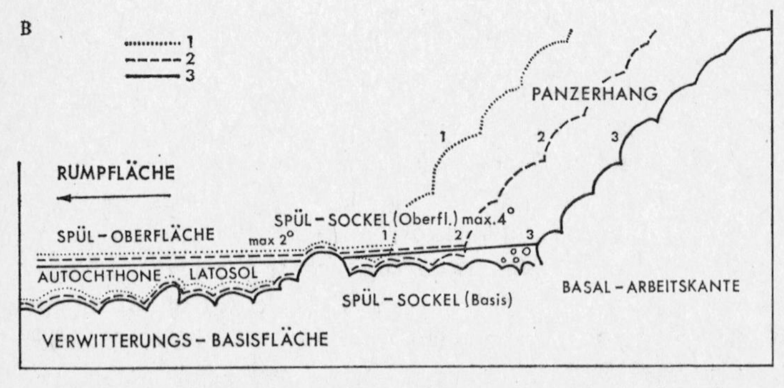

Figure 40 Formation and extension of tropical planation surfaces (after Büdel 1957, 1970).

A. Double levelling plain showing the formation of Rim wash-pediments (*Rand Spülpedimente*), shield inselbergs (*Schildinselberge*) and zonal inselbergs as a result of a lowering of the upper wash surface; B. Retreat of Panzerhang or 'armour slope' by basal sapping processes (*Basal-Arbeitskante*), with simultaneous lowering of upper wash surface (*Spül-Oberfläche*) and basal surface of weathering (*Verwitterungs-Basisfläche*) form *Rumpftreppen*. It is emphasised that such *Rumpftreppen*, or staircases of rim pediments, are not indicative of extensive rock planation, for in the intervening areas there is a simultaneous lowering of both the basal surface and the wash surface.

During the slow degradation of the weathered surface, humps of more resistant rock may emerge as isolated 'pediments' or low inselbergs: these are the peaks of the irregular basal surface of weathering. Once exposed, such rock surfaces are climatically arid and clay weathering gives way to block disintegration. So long as powerful sheetwash continues to remove this debris, a marked break of slope will persist between the exposed rock and the contiguous weathered surface. Inselbergs found clustered near the mountain rim (*Auslieger*, or outlier, inselbergs), however, are formed by scarp retreat and have no connection with the azonal or 'shield' inselbergs of the plains. These hills may be very high and may carry remnants of former ancient surfaces on their summits. As the surrounding wash surface is lowered, so the shield inselbergs may grow in height, but they remain as isolated outcrops unless rapid stripping of weathered debris supervenes.

Rapid stripping, according to Büdel, may result from tectonic uplift, or from a shift towards a more arid climate. To expose the uneven basal surface in its original form, stripping must be rapid, and will reveal an uneven *Grundhockerrelief*. This is seen to have occurred in the southern Saharan areas such as the Adrar des Iforas, where an uneven inselberg topography is interpreted as a relict of former humid climatic systems. The major effects of increased aridity are to slow down deep weathering and increase surface erosion. Streams may receive a coarse load, but are generally overloaded with debris which is mainly used for lateral planation and the extension of pediments. If removal of the weathered layer is slow, then modification of the basal surface must occur, resulting in the formation of extensive rock plains (*Fels-Rumpflächen*) within which the shield inselbergs do not grow in height but are continually lowered with the surrounding plains.

Büdel thought that, whilst the double surfaces of levelling continue to control relief on many of the high plateaux of Africa, most recently uplifted surfaces are in fact dissected into valleys. This destruction of tropical plains is most marked in the semi-arid and temperate zones. An important part of Büdel's reasoning applies to central Europe where remnants of such development are seen in the uneven summit areas of European mountains, and in rim pediments formerly thought to be relics of once very extensive planation surfaces.

This comprehensive hypothesis of tropical relief development has provided a framework for much recent thought, and has emphasised many of the principal characteristics of terrains developed over tropical shields. It integrates work on deep weathering with concepts of differing erosional mobility between rock and regolith. However, like most general theories, it pays little attention to the detail and variety of tropical landscapes.

Birot (1968) placed similar emphasis upon the deep decomposition of crystalline rocks, and the 'paralysis of linear erosion', leading to an abrupt transition from one surface to another across steep escarpment zones, but he recognised also the fundamental fragility of deeply weathered terrains. This fragility can lead to rapid dissection, even in conditions of relatively weak stream erosion. Slight tilting or uplift will result in destruction of old surfaces which in detail are poorly preserved in the humid tropics, except where protected by lateritic duricrusts. It is perhaps significant that whilst Büdel (1957) grouped all seasonally humid tropical climatic zones as essentially one (figure 3, p. 6), Birot (1968) adopted a morphogenetic approach which separates the humid tropics from the semi-arid and arid zones. According to Birot it is in the latter zones that the development of plane surfaces and abrupt breaks of slope are most characteristic. These features are well documented by Peel (1966). On the other hand the much larger areas of savanna climate with alternating wet and dry seasons combine features of both extremes, together with the widespread occurrence of laterites. According to Birot (1968) this implies that pediments extended by efficient sheetwash become preserved by the development of laterites, whilst the abrupt piedment angles at the foot of residual hills are maintained by intense weathering due to abundant moisture supply at this level.

Neither of these writers explored in detail the possible effects of climatic oscillations, although both refer to an acceleration of stripping processes in response to increased aridity within the humid tropics. It is also probable that too great an emphasis has been placed on the supposed feebleness of channel erosion by tropical streams (see p. 120). Two fundamental questions arise immediately from this analysis. First, is it possible to interpret tropical landsurfaces according to varying degrees of deep weathering or stripping of regolith? Secondly, what relationships exist between these surfaces and the hillslopes which rise above them as inselbergs or escarpments?

Weathered and stripped landsurfaces (*etchplains*)

Mabbutt, in a series of studies from central and western Australia, has clearly distinguished between weathered and stripped landsurfaces (Mabbutt 1961a, 1962, 1965a). However, these analyses were concerned with semi-arid and arid landscapes in which the deeply weathered landsurface was regarded as inherited from a more humid period, responsible also for the widespread development of duricrust which overlies and protects the old weathering profiles. It is significant that climatic change becomes central to many arguments concerning the apparent alternation of weathering and stripping in the history of landscapes.

Mabbutt commented on the occurrence of tor studded plains which 'appear to be etchplains, under which interpretation on the stripped landsurface should reproduce the form of the prior weathering front' (1965, p. 99). Thornbury (1954) did not consider that etchplains of wide extent were likely to develop, and Bishop (1966) has discussed the use of the term in type areas described by Wayland. Bishop (1966, p. 149) observed that 'there is no stripping to expose considerable areas of the basal surface of weathering as a result of differential removal of the soft rotted regolith and with accompanying formation of extensive plains', and he concluded that 'the fact that the peneplains and pediplains of Uganda are cut across deeply weathered rock which in places attains 200 ft or more in thickness, militates against the use of *etching* to describe the mode of formation of extensive areas of low relief' (p. 149).

However, I do not personally share this narrow view of etchplanation. The sense of the term 'etch' is to corrode a surface by chemical means, and this clearly is the process affecting the rock surface at the weathering front, whether this is viewed on the broad scale as an undulating rock surface varying in depth below the land surface, or on a micro scale as applied to the surface of a single core boulder. Nomenclature for such landscapes is not particularly important, but it should be helpful both for description and interpretation. The distinction between weathered and stripped landsurfaces becomes difficult where this twofold division is inadequate to account for the variety of geomorphic surfaces present in an area where the twin processes of etching and stripping have been or continue to be, prevalent.

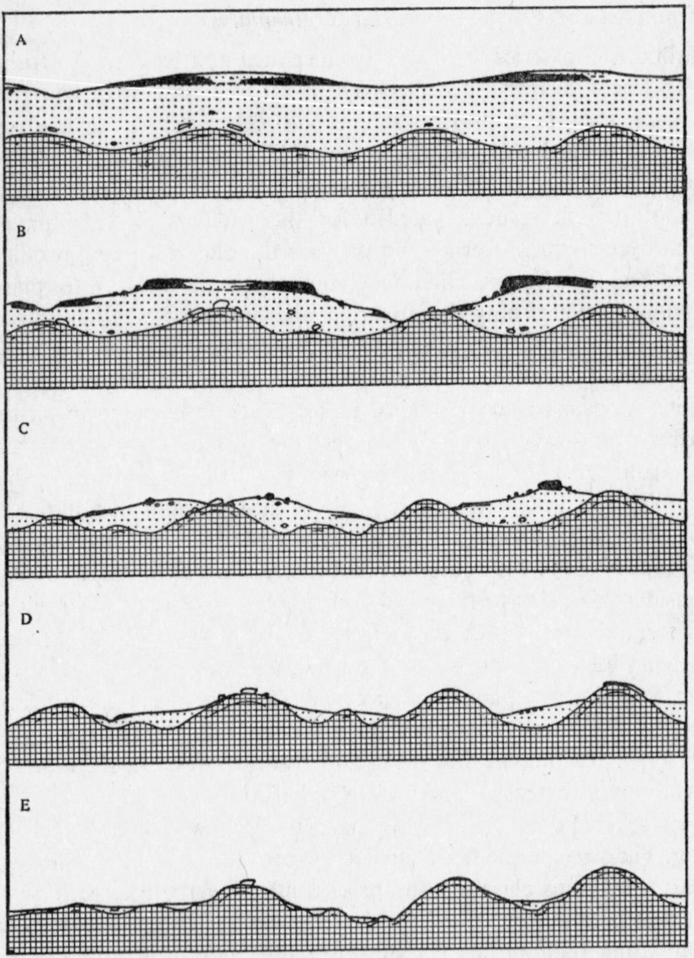

Figure 41 Principal types of etchplain, and etchsurface over crystalline rocks*.

* In the text two such types are recognised: incised etchsurfaces (E) and pedimented etchsurfaces (F). It is doubtful if these can be separately identified without ambiguity and they are shown together in the diagram.

A classification of etchplains and etchsurfaces

A scheme for the elaboration of the etchplain concept was therefore advanced (Thomas 1965b, figure 41) in an attempt to overcome this difficulty. Six principal types of etchplain or etchsurface were described in which varying degrees of stripping from a lateritised plain were envisaged. Effects of incision into the basal surface of weathering, and the extension of pediments, can also be recognised (Thomas 1969). These are illustrated in the accompanying diagram (figure 41):

A. *Lateritised etchplains* are surfaces of low relief underlain by extensive laterite deposits. Ideally they should exhibit a few signs of recent steam incision, and should carry few, if any, residual hills above their general level. Most examples, however, exhibit some degree of steam incision below the laterite horizon, usually as a result either of recent warping or of climatic and vegetational changes of the Pleistocene.

B. *Dissected etchplains* are therefore recognised as the first stage in modification of lateritised etchplains, following accelerated stream erosion, and resulting in the formation of duricrust breakaways overlooking well-defined valleys between gently sloping tablelands. Exposures of the basal surface of weathering in the form of tors or domes are unusual except in localised areas of shallow weathering, but rock may be exposed extensively in the more important stream channels.

C. *Partially stripped etchplains* follow further dissection and stripping of the deep regolith. At this stage much of the former laterite cover will have been removed; a few mesas may survive, but most laterite deposits will be of a secondary, concretionary type. Rock outcrops in the form of tors and low domes may be common, and occasional bornhardts may appear. Over wide areas soil development will be on truncated weathering profiles: from the mottled, pallid or transitional zones containing corestones and unweathered rock fragments.

D. *Dominantly stripped etchplains and etchsurfaces* represent a further development of the stripping process, when all but the deepest basins of weathering and areas of least resistant rock have become stripped to form a complex, multi-convex surface of exposed rock and shallow regolith, carrying important areas of lithosolic soils. Where the basal surface is markedly uneven this type of terrain merits the term, *etchsurface* rather than etchplain. Such stripped surfaces are usually a result of important tectonic warping, or are a response to climatic changes, especially in the savannas. Localised occurrences of this type of relief coincide with particularly resistant rock types such as migmatitic acid gneiss.

E. *Incised etchsurfaces* occur where the basal surface of weathering has not simply been exposed but has become subject to widespread modification by steam erosion. This will frequently occur in conjunction with dominant stripping and as a result of important base-level changes.

Climatic oscillations will seldom be competent to produce important modifications of the basal surface (see Mabbutt 1961a). In many instances the incised etchsurface is absent or only developed over small areas, and the stripped etchplains abut against true escarpments, formed either as a result of differential erosion or because of tectonic upheaval and critical height failure.

F. *Pedimented etchplains and etchsurfaces* evolve from stripped or incised etchplain types, when stream incision ceases and valley sides undergo slope retreat beneath shallow weathering profiles (Jessen 1936; Ruxton and Berry 1961a). It is doubtful whether this type of landform system can be distinguished from the stripped and incised types.

Ollier (1969) has criticised this scheme on the basis of Wayland's original definition and has suggested that the terms: *deeply weathered plain, partial etchplain,* and *etchplain* be used for the stripping sequence. He also suggested that the term *complex etchplain* might be used for surfaces over which both weathering and stripping continue contemporaneously. This problem of nomenclature is partly a question of viewpoint. Personally, I take the view expressed by Büdel (1957) that the fundamental denudation (etching) occurs along the basal surface of weathering. As this etched surface is exhumed by processes leading to stripping of the regolith, so it may appear as a dominantly stripped etchsurface (*Grundhockerrelief*) or may be modified by surface denudation to become a pedimented etchsurface. The sense of Wayland's argument was that the balance of processes, leading either to weathered or stripped surfaces, will alternate with the rate of tectonic upheaval, to which must be added the important influence of climatic variations. Etching is most active during periods of landsurface stability, when surface denudation operates mainly across the regolith and the basal surface is etched into hummocks and hollows, ridges and grooves, by weathering. The instability of landsurface which leads to accelerated stripping also exhumes at least the upper rises of the etched, basal surface, but at the same time the relief becomes more diversified. It is clear from most records (Mabbutt 1961a; Ollier 1965; Thomas 1966a) that the basal surface is not continuously exposed over very wide areas, and that regolith is retained in depressions, often protected by thresholds of unweathered rock (Ollier 1965). If, therefore, the term etchplain is to be retained, it should logically, by use of qualifying terms be extended to cover the range of landsurfaces with which etching and stripping are associated.

The alternative is to regard these landscapes as varieties of pediplain, and this view is implicitly or explicitly taken by many writers. The term etchplain is preferred here in so far as it distinguishes the landsurfaces under discussion from those which evolve by rock planation. However, since the extension of 'slope pediments' within the regolith is a major part of the stripping process, these surfaces are, in a certain sense of the term, pediplains. The two concepts should not therefore be seen in opposition and choice of terminology may depend upon individual emphasis. German geomorphologists refer almost exclusively to pediplanation, whilst the Dutch writers Bakker and Levelt (1964) have tended to equate the two terms.

A more serious criticism of the concept of etching and stripping might be the continuing emphasis upon erosional forms, at the expense of enquiry into patterns of deposition which have been shown (Chapter 5) to be of very great importance to our understanding of tropical landscapes. In the study of tropical soils, translocated materials are no less important than patterns of deep weathering and outcrop, and in the analysis of the process of stripping it must be realised that much of the material removed from interfluves, accumulates at lower levels in the landscape (Fölster 1964, 1969). This point will be explored further, particularly in respect of evidence for climatic change.

If the etchplain concept is to be of more than academic value (and to have this it must be genetically sound) it must fulfil several conditions: it must express the dominant characteristics of the landsurfaces involved; it should be capable of quantitative expression which will permit some more objective method for the designation of particular terrains as types of etchplain, and it should be capable of further extension and modification as new facts emerge or new landscapes are examined. And last, to be of practical use the types of etchplain should express characteristics of terrain which have meaning for cognate studies of soils and foundation materials.

In fact it matters little what terminology is used and many would advocate a purely quantitative basis for description which avoids all problems of interpretation. However, if we can summarise a wide range of land properties by the use of such terms as 'lateritised etchplain' this may assist in the logical subdivision of terrain. The terrain types illustrated in chapter 6 (table 14, p. 162) may be described respectively as dissected and dominantly stripped etchplains. Although no attempt has here been made to provide a quantitative expression for the morphometry of each etchplain type, such an attempt would be difficult, because of the variety of local conditions, and lack of measured examples which can be used for this purpose. It is certain that each broad class suggested here would require to have many local forms, but it is necessary that the definitive properties should include the occurrence of particular surface materials such as lateritic duricrusts or bare rock surfaces. Such an approach corresponds closely with Moss's classification of soil catenas (Moss 1968; Thomas 1969).

ETCHPLANATION AND PEDIPLANATION

Büdel (1957) described deeply weathered plains bordered by rim pediments, and similar descriptions of tropical landsurfaces have come from Ollier (1960), Mabbutt (1961a) and Thomas (1969). Jessen (1936) on the other hand emphasised the extension of plains as a result of basal sapping and the subsequent weathering of the extended plains (figure 42). Although the processes invoked are different, this reasoning is similar to that used by many other writers including Dixey (1955) and King (1953, 1962), who have described such a development as pediplanation. Perhaps the clearest exposition of the situation has come from Ruxton and Berry (1961a), who relate the hillslope development not only to climate, but also to the existence of adequate relief development and the possibility of a two storey landscape.

In Wayland's model of the etchplain elaborated here such a development is not envisaged: continuous or discontinuous rise and tilting of the landsurface was not such as to generate escarpments from tectonic causes. Incision of streams was generally restricted to the depth of the regolith, or the processes of stream erosion were roughly paced by continued weathering. Residual hills and local escarpments developed as a result of differential weathering and erosion, for instance of gneiss against granite. As Jessen's diagrams show (figure 42) it is the exposure of the fresh rock surface along the escarpment that fundamentally alters the morphogenetic framework.

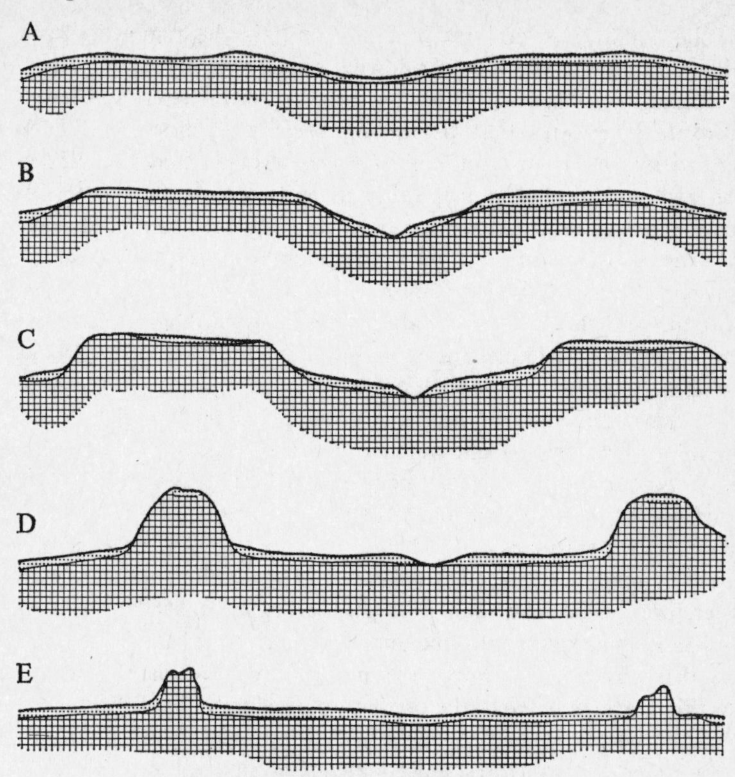

Figure 42 Development of inselberg landscapes by the extension of pediments on crystalline rocks in the savanna zone (after Jessen 1936).

One problem which affects much of this discussion is petrographic, for as Bakker and Levelt pointed out (1964), many theories relate to crystalline rocks or even specifically to granites. Another problem is climatic, for the assumption that evolutionary schemes for the tropics may be applied uniformly throughout the zone or result from monoclimatic histories may be grossly misleading. These points are considered further.

The generation of escarpments or hillslopes which exceed in height local depths of weathering (other than by differential erosion) is in some ways a separate issue from the question of their subsequent evolution. According to Jessen (1936), Freise (1938), Wahrhaftig (1965) and Hurault (1967) the critical factor in the evolution of both hillslopes and residual inselbergs is the removal of fine debris from the hillslope. Following the emergence of bare rock surfaces the differential rates of weathering on exposed and buried rock faces ensures that weathering becomes concentrated below the rock face. However, not all hillslopes possess extensive outcrops of fresh rock. In humid tropical areas particularly, the rock surface over granites remains covered by a thin regolith, commonly containing corestones. Footslopes below such hillslopes are commonly deeply weathered in the vicinity of the escarpment, as shown by Ruxton and Berry (1961a) and by Clayton (1958) and a piedmont, rock bench or pediment is therefore not always evident. In fact within the humid tropics such occurrences are rare, but in the semi-arid, savanna lands of northern Nigeria, stripped rock surfaces containing vestiges of old weathering crusts extend from the footslopes of escarpments in certain places (Plates 25, 26). Similar rock surfaces which display evidence of recent stripping are found elsewhere in the savannas.

The climatic and vegetational control over such phenomena may therefore be decisive, and it might be argued that the stripping process, as a widespread phenomenon, is confined in the main to the drier savannas. However, two points need to be made in respect of this suggestion. First, the oscillation of climatic zones during the Pleistocene period have brought such conditions to areas that are now either desert or forest environments. Secondly, it can be argued that the stripping process is accelerated by the *disturbance* of established ecosystems, and may not therefore be associated with any single climatic zone. These points are explored further in chapter 10.

Studies of footslopes in humid areas seem to confirm that they are commonly deeply weathered. There seems to be little justification, therefore, for any suggestion that these slopes are rock cut pediments. The deep weathering is an essential part of the morphogenetic system, which also affects the hillslope and probably leads to its retreat. Slope retreat and etchplanation are possibly linked in a similar way to slope retreat and pediplanation which appears to be a separate system, characteristic of semi-arid areas.

A pediment containing regime must of necessity be one of lateral planation, whether by streams or by weathering processes, and a criterion for pediplanation might possibly be the ability of the denudational system to remove the products of weathering from the base of the hillslope and to use these in the fashioning of a thinly veneered, rock surface. This is contrary to Clayton's view (1956) in which the relative absence of weathering from *upper* hillslopes was seen as the main criterion. Certain thresholds in the morphogenetic system clearly exist which may affect the balance between accumulation and removal of material at any given point in the landscape. In homogeneous terrains, preferential accumulation on steep slopes or on upper facets of hillslopes is most unlikely so that basal attack of the hillslopes and some kind of retreat are almost inevitable, except in circumstances where the base of the slope is continually being lowered. This may occur in areas of rapid stream erosion and also where the hillslope marks a contact between rocks of markedly differing resistance to the denudation system.

If the processes of etching can be construed equally to attack in the lateral plane (figures 39, 40), etchplanation may be regarded as the primary mode of landform development, even where escarpments separate upper and lower landscapes. In fact the weathering system along the scarp front can only be understood if lateral and vertical encroachment of the weathering front are considered together.

The conflict of views between those who describe all tropical plains as pediplains and those who emphasise the role of deep weathering is clearly illustrated by two recent studies of parts of Guyana (McConnell 1968; Eden 1971). To a considerable extent the difference in interpretation can be explained in terms of the scale of enquiry and the emphasis placed upon individual features of the terrain. To McConnell (1968) the existence of stepped planation surfaces, separated by steep escarpments, is clear evidence of pediplanation, whilst to Eden (1971) the relative scarcity of flanking, rock pediments around residual hills, and the extensive deep and lateritic weathering, together with the presence of rock landforms associated with the stripping process, are indications of effective weathering processes and the development of etchplains.

Notwithstanding these reservations, it is clear that the retreat of major hillslopes does take place within the humid tropics, but this occurs in response to basal weathering attack. However, the diversity of relief on the plains which make up most of the landscape, can only be interpreted in terms of the interplay of deep weathering, stream incision and slope processes taking place mainly within the regolith (figure 43). It seems to me that such plains are better described as etchplains since this term does not imply a particular mode of slope development.

Etchsurfaces and lithology

This simple model of etchplain development takes no account of the effects of lithology. But in order to account for the diversity of landforms over extensive areas, the effects of rock character must be explored. The concept of differential erosion is as old as the science of landform study, but its application to tropical landscapes requires a different emphasis from that commonly used in a more general context. If denudation begins by rock decomposition along a weathering front, rather than by abrasion of rock surfaces, then differential weathering must be considered the primary cause of differences in elevation and morphology between terrains developed on various rocks—at least on old lands.

On tropical shields the tendency for rocks rich in ferro–magnesian minerals to weather rapidly and to form thick lateritic mantles is undisputed. Areas of basic igneous rocks and metamorphic rocks such as biotite schists and amphibolites commonly carry a thick duricrust which protects the underlying clays from erosion. As a result, they generally form areas of positive relief, and may rise as plateaux above more 'resistant' rocks such as granite or acid gneiss. This is seen clearly in central Sierra Leone, where the Kambui schists form a series of elongate uplands overlooking gneissic terrains. It is commonly observed that these, more acidic rocks, whilst they may exhibit local deep weathering, commonly do not carry extensive laterite sheets. Observations of this phenomenon are common and come from Africa and from South America (Bleackley 1964). Consequently the regolith forming over granitic rocks is liable to rapid removal from exposed sites. This is a major reason for the occurrence of inselberg landscapes over these rock types, and for the limitation of dominantly stripped surfaces to areas of more acidic rocks. On the other hand deeply weathered granites and gneisses are found, so that it is not a corollary of this observation that such areas were never deeply decomposed. Furthermore, laterites are found over acidic rocks in places, especially where lateral movement of iron from neighbouring formations has apparently taken place.

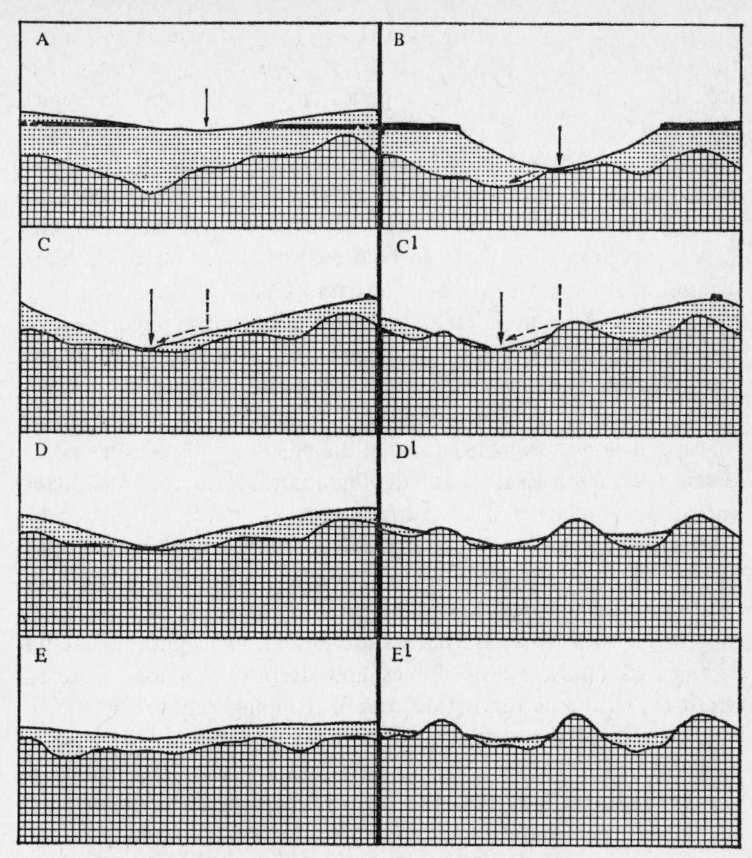

Figure 43 Alternative modes of valley development over weathered crystalline rocks in the tropics.

A. Lateritised etchplain or weathered landsurface; B. Dissection leading to 'superimposition' of stream channel on basal surface of weathering (dissected etchplain); C, D, E. Simultaneous lowering of landsurface and weathering front without progressive stripping. Common in the rain forest zone and over less resistant rocks.

C¹, D¹, E¹. Progressive stripping to form stripped etchsurfaces. At stage C/C¹ the stream channel shifts towards troughs in the basal surface of weathering.

Over the African shield where the country rock around granitic intrusions is generally gneissic, the granites usually stand out in bold relief, either as elevated plateaux or as groups of domed inselbergs (figure 30, p. 169). But where granites have been intruded into less weatherable sedimentary strata, as in eastern Australia, they frequently occur as embayments or lowlands between sedimentary ranges. Sedimentary rocks overlaying the Basement Complex in Africa also commonly form bold escarpments. One of the most striking of these is the Voltaian Scarp in Ghana (Hunter and Hayward 1971). Also in Ghana, the metasedimentary phyllites form striking hill ranges, commonly capped by lateritic bauxites, and quartzites nearly always stand out as prominent ridges.

To encompass these variations within the etchplain or etchsurface concept, it is necessary to envisage the effects of dissection upon a terrain weathered to different depths according to rock character, but as yet undissected by surface erosion. Under a favourable climate and vegetation cover, lateritic formations will have developed, particularly over the basic rocks. Dissection of such a terrain will create breakaways in the laterites, and may give rise to rapid stripping of regolith where duricrusts are absent. Over granites a compartmented landscape of domes and deep valleys may develop, whilst over gneisses varying degrees of stripping may occur.

The onset of the stripping process may be induced by climatic changes, or it may, as Wayland stated, be tectonic. The pattern of regolith removal may take the form of widespread denudation of watersheds as a result of increased overland flow, or may radiate from the principal river valleys as they become incised and valley-side slopes are steepened. The result is to bring into being dissected, stripped and incised etchsurfaces according to position in respect of the drainage net, and according to the nature of the pre-weathering of the underlying rocks.

The deeper the dissection the greater become the contrasts between areas of different lithology or mineralogy. This principle is well illustrated in central Sierra Leone, where local relief rises to at least 200 m, within and around the Sula Mountains. This area receives a high but seasonal rainfall of about 3000 mm per year and the climate is clearly conducive to lateritisation. Very deep weathering and a thick duricrust overlies the schists, which are dissected into deep valleys with bounding slopes exceeding 30° in many places. Mass movement on these slopes has resulted in many slip scars and earthflows which contain massive blocks of duricrust recemented on the steep slopes (figure 18, p. 80). To the west the Gbenge Hills, a series of striking granite domes, rise almost to the height of the adjacent plateau, but the surface at their base, which also extends across surrounding gneissic terrain is at a much lower level. Over the gneissic plateau to the east are accumulations of colluvial lateritic rubble that have probably resulted from the dissection and reworking of a thinner duricrust as the gneissic surface has been lowered. Differences in topographic level over the gneiss terrain suggest the operation of more than one cycle of planation, and the conclusion is unavoidable that repeated lowering, involving partial stripping and reweathering of the gneisses, has occurred throughout a long period during which the lateritic uplands over the schists have survived as an elevated plateau.

The evolution of these landscapes has clearly involved the operation of several periods of planation, probably according to the broad outline offered by King (1962). In fact we have little knowledge of the initiation of the present sub-aerial landscape over wide areas. It is possible to point to broad areas of gently undulating terrain, apparently developed across a variety of rocks in certain areas, but elsewhere a diversified relief may have persisted for a long period. The relationships between lateritised etchsurfaces and stripped surfaces carrying inselbergs suggest that, whilst periodic alternations between weathering and stripping have effectively lowered the latter, the lateritic terrains have been marginally eroded, but otherwise little affected.

The accentuation of relief through differential weathering and erosion may give rise to lithologically controlled escarpments, and to belts of dissected country, exhibiting a strong development of ridges and valleys, as in the forest zone of south-western Nigeria, where north–south lineation in the Basement Complex has been exploited by south flowing streams.

Differential weathering and erosion may also exploit apparently minor variations in rock composition, texture or jointing. Within granite, for instance, enclosed or semi-enclosed basins occur at different scales. Thorp (1967a) noted these features within the granite Ring Complexes of northern Nigeria, where they respond to different phases of granite emplacement and also to internal variations in jointing frequency within a single granite. Basin features within granite terrain often occur in relation to weathering patterns (see Chapter 3 and figures 20, 21) and tor landscapes (Waters 1957; and chapter 7). It has been noted that many such basins remain floored by weathered material, but rock-floored basins containing tors and corestones on lower slopes are found in many areas, such as the Monaro of New South Wales and New Zealand (Thomas 1974). Some of these basins are totally enclosed and cannot therefore have been formed by fluvial erosion which normally links the basins across thresholds or via narrow exit valleys. Outside of the glaciated zone such basins must have arisen through eolian action or by solutional removal of weathering products. This last hypothesis is hinted at by Twidale (1971), and although Ollier (1967) has argued for constant volume weathering, at certain stages in the weathering process a reduction in volume in the rock material may take place, and in the absence of strong fluvial erosion shallow basins might be formed in this way.

Whatever the detailed explanation, such basin landscapes appear to mark areas and periods of widespread stripping of regolith, perhaps due to major bioclimatic disturbance. The basin morphology with its strong concavity of form also appears to mark periods of lateral planation, in contrast to the formation of multi-convex relief in weathered granites, where vertical stream incision is evident.

On the basis of detailed studies of granite terrain in French Guiana and West Africa, Hurault (1967) has proposed three main systems of denudation according to zonal climate and vegetation. Within the rain forest (where rainfall exceeds 1800–2000mm), Hurault noted the predominance of cupola-shaped or 'alveolate' relief with low internal relief over wide areas. This type of surface is deeply dissected along continuous fractures in areas of strong regressive erosion and the cupolas develop into high domes, especially where the basal surface is exposed by landslides. This type of development corresponds closely with the morphology of granite terrain in southern Nigeria (figures 30, 35) and in Sierra Leone.

In contrast, Hurault claims that regressive erosion in the savannas (where rainfall varies from 1300–1600 mm) proceeds by the recession of escarpments whilst the alveolate landscape becomes re-established below the escarpment front. An intermediate situation was recognised, where rainfall amounts total 1500–1800 mm, within which lineal erosion predominates, but is propagated upstream via steep valley heads which lead to the excavation of domes. These conclusions were arrived at from a study of many thousands of aerial photographs, but no morphometric details were obtained. Moreover the clear statement of morphogenetic zones based on contemporary rainfall regimes does not take account of past climates, under which many of the forms described must have evolved. Experience in Nigeria suggests that both ridge fronts and dissected terrain may be found over a wide range of climates.

The case for closely defined morphogenetic zones to account for major relief forms is difficult to establish, although broad distinctions, particularly in the character of footslopes (figure 39) can be made. Attention in a subsequent section will accordingly be devoted to the influence of regional climates and climatic oscillations upon the detailed forms and deposits of tropical landsurfaces which offer a fuller record of morphogenetic events.

Nevertheless the manner of land surface lowering in tropical landscapes becomes a matter of some importance to the present arguments. The retreat of escarpments has been ascribed largely to basal sapping processes, and the fundamental lowering of plains to etching along the weathering front, followed by stream incision and the extension of pediments mainly within the weathering profile. According to this view, denudation results in the replacement of one set of forms by another, and does not involve the systematic lowering of the landsurface without change of form.

Trendall's hypothesis

Trendall in 1962 suggested, however, that steady state forms might develop in lateritised landscapes. His ideas were considered (p. 67) in relation to the development of laterite deposits, and it was concluded that many of the premises in his analysis could not be accepted. His basic hypothesis was that a lateritised landsurface with moderate internal relief would become lowered without substantial changes of form, as a result of solutional removal of material from the underlying pallid zone. However, although such a scheme may not be acceptable in the form advanced, it still requires further consideration. One reason for this is the importance accorded to solution by many geomorphologists. Rapp (1960) found that solution exceeded all other processes in magnitude, even in a periglacial environment, whilst Ruxton (1958) found solutional losses in the Sudanese savannas amounted to more than 30 per cent of original rock material in some cases. It is therefore an attractive proposition to suggest that this forms not only a major component of mineral loss in rocks, but also a major part of the total denudation, especially in the tropics. However, such an hypothesis encounters fundamental difficulties. Some of those associated with laterite formation have been considered (p. 68); those which affect concepts of landscape lowering can be summarised:

(1) As Ollier (1969) pointed out, the underlying pallid zone from which material must be removed in solution, commonly preserves details of the rock fabric. This is a sign of constant volume alteration (Ollier 1967), and must imply that surface lowering is almost wholly confined to surface erosion.

(2) The development of breakaways as a result of stream incision, which forms a part of Trendall's hypothesis, undoubtedly leads to lateral slope retreat. Evidence for this is found in the form of redistributed laterite gravel and other slope deposits found on the surrounding pediments (Ruhe 1956; Moss 1965; Fölster 1969). In this way, and as a result of the process outlined in (3), lower slope laterites develop, and they clearly do so at the expense of the upper laterite sheets. So widespread is this relationship that a general hypothesis requiring the replacement of one set of forms by another appears far more acceptable than the alternative of continuous development of a single landsurface.

(3) The development of relief within lateritised terrain will induce lateral water movement within the upper zones of the regolith. This throughflow removes iron from within and below existing duricrusts and leads to its concentration and precipitation in other locations, commonly on lower slopes of valleys, particularly where convergent flow occurs (Fölster 1969).

(4) Field situations offer evidence of localised lowering and slumping of duricrusts, but seldom point to a general lowering of the landsurface. Thus, in the Sula Mountains under conditions of heavy rainfall, local slumping and widespread cambering of the duricrust is evident (figure 18, p. 80), but the lateritised surface remains elevated above neighbouring plains. Because the degree of weathering represented by the laterite cover is much greater than that

encountered over neighbouring, acidic rocks, it is difficult to argue that the latter will have been *differentially* lowered by solutional processes. The conclusion in this and many similar cases must surely be that weathering profile is preserved by the laterite crust at or near its original elevation, until the crust itself is dismantled by marginal attrition or by mass movement. Most laterite caps display slight tilting which may reflect the attitude of the original lateritic soil horizon, but in some cases it may reflect local as opposed to wholesale lowering of the deposit. In other cases the observations concerning the occurrence of considerable internal relief in lateritised terrain may be due to an assumption that the deposits developed across the terrain are all a part of a single laterite. Brosch (1970) has examined the laterites in south-eastern Uganda and has concluded that the relief is made up of several different laterites, many of secondary origin and some of very recent formation.

(5) The mechanism by which lowering could take place is not clear. Trendall (1962) considered that this would be mainly by removal of solutes from below the cap. But Ollier (1969) has pointed out that in a fully developed laterite profile this would require the removal of material from the pallid zone which, by definition, is already strongly leached. De Swardt (1964) in support of the general hypothesis suggested that lowering of the surface would be achieved by 'solution in depth and physical erosion at the surface'. But it is difficult to see how such a mechanism could work since erosion of the surface would inevitably be at the expense of the laterite deposit, leading, as already observed, to the redistribution of laterite fragments across pediment slopes.

There remains, however, a possibility that, within the rain forest where the laterite remains as a soil horizon and is not indurated, dissolution of the iron *in the laterite* may occur, permitting the gradual lowering of the deposit, during periods of relief development. This could occur without the need for volume alterations within the mantle below the laterite, and would imply only that topsoil removed from the profile would be replaced by renewed weathering as the lateritised horizon moved downwards. Some of the evidence cited above argues against such a process as a widespread phenomenon, but it may contribute to the explanation of lateritised terrains.

It is possible to conclude, therefore, that neither the formation of laterite deposits nor the landforms found in duricrusted terrain require a developmental hypothesis such that as advanced by Trendall (1962). Furthermore, other evidence, particularly from translocated materials on slopes, does not support the concept of the survival over long periods of steady state landforms.

However, even where replacement of slopes occurs as laterite sheets are broken up along valley flanks and around mesas, a sequence of changes in the landscape may take place which require a minimum of morphological change (figure 43, C, D, E). Thus, over suitable rocks and during a period of slow degradation of the landsurface, removal of duricrusts may be accompanied by renewed lowering of the weathering front, in such a way that effective stripping does not take place. In an area of diverse lithology this may occur over less resistant rock formations, while stripped etchsurfaces are developed over adjacent rocks (figure 43; and Thomas 1966a). This style of development may help to account for contrasts in weathering and outcrop patterns in areas where relief development is slight and variations in relief and slope between different rock formations small. Such a situation appears to exist in south-western Nigeria (Thomas 1965b, 1969).

DEPOSITIONAL PATTERNS ON ETCHSURFACES

The processes leading to the stripping of regoliths, transfer sediment from interfluves and from valley-side slopes towards lower levels in the landscape, and eventually into stream channels. Study of semi-arid environments in Australia has led to the realisation that redistributed materials from old weathering profiles comprise a significant proportion of the total land area (Stephens 1946; Mulcahy 1961; Butler 1959, 1967), and in south-western Australia in particular, depositional patterns have been related to source horizons within sedentary regoliths (Churchward 1969). In fact the emphasis that has been placed upon the basal surface of weathering has tended to obscure the existence and importance of a basal surface of *erosion*, buried beneath superficial accumulations of gravel and hillwash which are commonly from 2–3 m in thickness, though they may exceed this locally.

Oscillations of climate have also been inferred from evidence of soil stratigraphy, and cycles of soil formation and erosion (K-

Cycles: see page 125) defined for parts of Australia. Ruhe (1956) emphasised the importance of transported mantles in central Africa, and recent studies by Fölster (1964, 1969) from the Sudan and from south-western Nigeria have confirmed the presence within the wetter savannas and the rain forest of extensive slope deposits. It has long been known that such deposits were widespread within the drier areas, where they include eolian material in the form of fossil dune systems, as well as alluvial fills and fan deposits.

De Swardt (1946, 1949, 1953, 1964), in a series of papers has pointed to common features of many stream valleys in tropical Africa. These contain apparently *in situ* laterite formations on interfluves, and secondary laterite deposits cementing transported materials on lower slopes. He described two major phases of valley incision and slope development which gave rise to the erosion of the summit laterites and later to the formation of bench like features from the secondary laterites which follow the present water courses. Moss (1965) has also emphasised the importance of colluvial transfer of residual material during the backwearing of laterite breakaways over sedimentary formations in southern Nigeria (figure 17).

Reference was previously made (p. 130) to the significance given by Brückner (1955), Fölster (1969) and Burke and Durotoye (1971) to stone lines and patterns of colluviation in the southern parts of Ghana and Nigeria. These authors differ in details of field description and interpretation but their work indicates a common pattern of one or more pediment gravels (stone lines), sometimes ferruginised and usually overlain by variable thicknesses of loamy sand, interpreted as hillwash (Fölster 1969), or to the effects of termites and earthworms. Although the area studied by Brückner (1955) has been reinterpreted by McCallien *et al.* (1964), who found evidence of both marine processes and active tectonism, the general importance of depositional materials is clearly established.

Fölster's (1969) work is important since it is seen in the context of overall landscape development, particularly in relation to climatic oscillations during the last 30 000 years. He interpreted the alternate translocation and weathering of these materials in terms of short-term cycles of instability and stability of the landsurface. Such cycles appear to have been of general occurrence throughout much of the inter-tropical as well as the sub-tropical zone. Fölster regarded the slope deposits as cyclic features, resulting from shallow incision and periods of slope retreat or pedimentation within the soil profile. Parallel with Churchward's (1969) study in Australia, Fölster pointed out that 'depending on the depth at which the new erosion surface cut the former soil cover, the newly forming soil will contain relict features inherited from the older soil' (1969, p. 4). He also considered that environmental conditions will have varied markedly from those prevalent today. He developed a model for the retreat of low scarps within the weathering profile, during which a coarse gravel is left covering the basal surface of erosion, and, at a later stage, a fine grained cover or 'hillwash' is deposited over this gravel to complete the characteristic profile. Weathering and soil formation have subsequently affected this hillwash or 'pedisediment'. In many sections a repetition of these events is described using a nomenclature comparable with that of the Australian K-Cycle. Three or more such cycles are widespread. The climatic implications of these are later explored further, but it is notable that Fölster regarded the present landsurfaces in southern Nigeria as being 'remarkably stable' and inferred that drier conditions must have been responsible for the unstable phases.

An important part of this study was the incorporation of the lateritic duricrusts into the scheme outlined. Like De Swardt (1953), Fölster recognised an *older laterite* capping on interfluves, and a *younger laterite* forming bench like features along the flanks of valleys. These lower crusts were found on pediments cut back into the older laterite deposits, and thus they overlie translocated material in many localities, as well as being comprised of recemented fragments from higher sheets. These younger laterites are located in drainage depressions, especially in headwater dells, and along the gently concave valley sides of first order streams which have indistinct channels or no channel at all.

Churchward's detailed study (1969) of depositional patterns in a dissected laterite terrain over Cretaceous sediments some 100 miles north of Perth in Western Australia is of particular interest since it confirms many of the general comments already made about such landscapes, and extends their application to an area of sedimentary rocks. The area receives a predominantly winter rainfall of about 550 mm, and takes the form of a dissected plateau with an internal relief of up to 150 metres. The landforms and deposits recognised by Churchward (1969) are summarised in figure 44 and

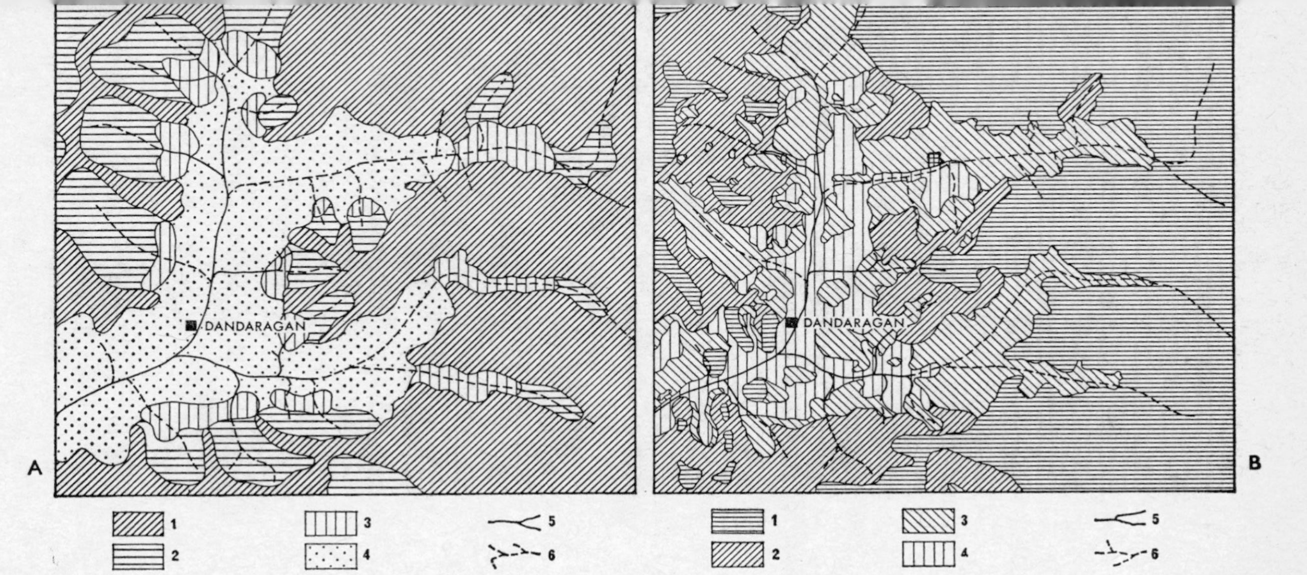

Figure 44 Landforms and deposits on a dissected, lateritised landscape in Western Australia (after Churchward 1969).

A. Topographic subdivisions
1. Undulating upland; 2. Upper tributary valleys; 3. Lower tributary valleys; 4. Main trunk valleys; 5. Perennial streams; 6. Valley floor with no stream course.

B. Deposits (see table 15).
1. Karamal; 2. Koodiwoodie; 3. Dandaragan; 4. Yere; 5 and 6 as in A.

in table 15. They indicate that material has been redistributed in a repeated pattern according to source horizons in the ancient weathering profile and the pattern of the erosional landforms. Deposition of the erosional debris was found to be highly localised, and in the trunk valleys accumulations were up to 16 m in thickness, suggesting to Churchward that the valleys had acted as local base levels rather than as channels for the evacuation of eroded material.

These two studies (Fölster 1969; Churchward 1969) both come from ancient shields, but in other respects the environments differ considerably. They indicate that under a certain range of conditions etchplains, developed by erosional modification of deeply weathered and lateritised landsurfaces, may also display characteristic depositional patterns. The provenance of the deposits on such surfaces will in turn reflect the available horizons of ancient weathering profiles in the vicinity. These will in turn be influenced by both the degree of antecedent stripping and the depth of dissection.

TABLE 15

THE RELATIONSHIP BETWEEN LANDSCAPE AND DEPOSITS NEAR PERTH, WESTERN AUSTRALIA

Landscape units	*Deposits*	*Source of deposits*
Undulating upland	Dominant: Karamal Minor: Koodiwoodie	Colluvium derived from the upper horizons of lateritic mantle
Upper tributary valleys	Dominant: Koodiwoodie Minor: Dandaragan	Colluvium derived from the deeper horizons of the laterite mantle
Lower tributary valleys	Dandaragan and Yere	Ferruginous sandstone Ferruginous sandstone and chalk
Main trunk valleys	Dominant: Dandaragan and Yere Minor: Koodiwoodie and Karamal	Indicated above Indicated above

(After Churchward 1969)

In Fölster's view the detail of the shield landscape of south-western Nigeria revealed periodic reworking of the upper zones of the regolith, in response to climatic changes. Minor pedimentation scarps could be induced by local as well as regional influences, leading to 'the fluctuation of the pedimentation impulse from one slope to another' (p. 28). The pediplanation process described, occurs normally within weathered rock; major irregularities of the basal surface of weathering leading to exposures of unweathered rock. During this process, the landsurface is repeatedly lowered by small increments, and Fölster comments that 'depth of weathering has been one of the main factors in the depth of lowering of pediments' (p. 47). This corroborates Mabbutt's observations (1961a) within a semi-arid environment, and suggests a general relationship arising from climatically induced stripping. Periodicity in hillslope erosion and deposition was also suggested by Mabbutt and Scott (1966), to explain complex profiles in coastal Papua.

10 Climatic Controls in Tropical Landform Development

The tropical climates embrace a wide range of rainfall regimes and vegetation communities, and there can be little doubt that the balance of geomorphic processes varies greatly from the perhumid rain forests, through the deciduous woodlands and savannas, and into the margins of the deserts. Yet some uncertainty persists concerning the effects of such variations, as they should be evident in the landforms. Throughout this study references have been made to possibilities of climatic control over such phenomena as weathering depths, landslide frequency, pediment development, valley form and depositional patterns. And it has been implied that these observations reflect the availability of water and the density of the vegetation cover amongst other controls. It has also been stressed that oscillations of climatic boundaries may have led to profound changes in the controlling factors at any particular location, such that the forms and materials that characterise the present landsurface may have been inherited in large measure from earlier periods of contrasting climate and vegetation. As Ruxton (1968c) has stressed, some aspects of the land complex adjust to changed conditions very much more quickly than others. Stream channel morphology and the extent of the drainage network probably adjust rapidly as climate and vegetation cover change, but it is quite clear that old weathering profiles and crusts survive long after the formative conditions have altered. The same may be true of aspects of the landform, such as slope and valley forms and residual hills.

Regional climates and patterns of climatic change

Attempts to define morphogenetic regions on the basis of contemporary climatic and vegetation patterns may therefore be quite erroneous, and we should perhaps attempt to view zonal landforms, if these exist, in terms of oscillatory patterns of climatic change. Such a view will be offered below, but it is relevant to begin with some consideration of the inner core of the humid tropics which according to Douglas (1969) can be interpreted as one of the few really stable ecosystems on earth. In equatorial lowlands, beyond the influence of changes in Quaternary sea levels, and in areas immune from recent tectonic disturbance, prolonged evolution of a rain forest ecosystem may have occurred, and the lack of seasonal periodicity in the equatorial climate emphasises the stability of the system.

Douglas (1969) pointed to the high productivity of the rain forest, which is more than twice that of temperate forests, leading to the plentiful supply of carbon dioxide and chelating agents to the soil. In this situation deep penetration of rock weathering is favoured, and many of the products of the weathering processes are re-cycled by the deep rooting trees. This is a good example of the condition of 'biostasy' advanced by Erhart (1955) in which Douglas (1969, p. 4) suggests 'increasing complexity of the plant community is associated with increasing soil development and a geomorphological trend towards a state of minimisation of work and even distribution of entropy throughout the drainage network. Under such conditions, most energy is used in the development of the plant community and not in erosion and removal of debris by streams.'

It is within these rain forest heartlands that we might expect to find a humid tropical denudation system (Douglas 1969) with particular denudational forms. But unfortunately much of this zone lies either in tectonically mobile areas or in recent depositional environments (figure 2, p. 5), so that representative areas of stable erosional landforms are restricted in occurrence. In fact the tectonic factor is of overriding importance in such a situation, and whilst tectonically active areas are restricted to the Indonesian region and the Andes, it is also clear that much of Africa and perhaps parts of Malaysia have experienced significant crustal warping or uplift that may have continued into the early Pleistocene. Positive movements of the crust will increase basin relief and slope values, disturbing the biostatic equilibrium, and may result in accelerated removal of weathered material. In the really active areas of New Guinea and Indonesia any trend towards biostasy is continually disturbed, so that rhexistatic conditions prevail (see p. 124).

Towards the margins of the rain forest, increased seasonality is reflected in a more open vegetation cover which under natural conditions grades into a savanna woodland in which seasonal leaf fall and annual growth of grass combine to expose the soil at the onset of the early rains. This seasonality can be expressed in terms of a greater concentration of rainfall* (Fournier 1962). But perhaps more important are the known fluctuations of humidity that have occurred in these climates, for there can be no doubt that the outer zone of the present humid tropics has experienced repeated desiccation during the Quaternary Era. The degree and frequency of such dry climatic phases and the nature of the more humid periods separating them is little known. But evidence of slope deposits (Brückner 1955; Fölster 1969), together with inferences from the distribution of duricrusts, and possibly inselbergs, all point to such events. Moreover, with increasing latitude the severity of the dry phases becomes more pronounced. In the northern savannas of West Africa evidence for former arid conditions comes from the great fossil dune fields (Grove 1958; Grove and Warren 1968), whilst more humid periods are marked by lacustrine and alluvial deposits, together with evidence for a much enlarged lake in the Chad Basin. This 'Mega-Chad' left shorelines as much as 40 m above the present water level.

It is unlikely that this increased humidity of climate can be traced solely to increased rainfall. Changes in solar radiation may have lowered temperatures within the tropical zone by several degrees, reducing evapotranspiration which may have been lessened further by greater cloudiness of climate. A recent review of ecological data from Africa by van Zinderen Bakker (1967) emphasises some major features of Quaternary climatic change and points to some of the difficulties of interpretation. Studies from East Africa (see Coetzee 1964) show that on Mount Kenya ericaceous vegetation extended to altitudes as much as 1000 m below present distributions at a date of around 18 000 B.P.,† during the Würm maximum of northern Europe. After 10 300 B.P. the forest zone again began to colonise to higher altitudes marking an amelioration of climate. In Angola

* Concentration as defined by Fournier can be expressed as p^2/P, where p^2 is the square of the maximum mean monthly rainfall, and P the mean annual rainfall.

† All dates are given in years before the present century (B.P.).

(7°36′S) at an altitude of 800 m, a core showed evergreen montane forest with a date of 38 000 B.P. This vegetation is today found above 1600 m and again indicates at least 850 m depression of the vegetation belts. Quoting Flöhn (1963), van Zinderen Bakker (1967) suggested that these shifts imply a fall in mean annual temperature of 0·51°C for every 100 m, amounting to a possible total reduction of 5·1°C during the Würm maximum from perhaps 27 000 to 10 000 B.P.

Walker (1970) has recently offered some comments on vegetational changes in New Guinea based on palynological findings. Sites between 2000–2800 m, within the present altitudinal limits of oak forests, indicated the presence of sub-alpine conditions between 26 000–22 000 B.P., and this again would require a depression of the vegetation belts by at least 850 metres.

Although chronological correlations are difficult to make in any detail, and much remains to be discovered about such changes, it is clear that a major depression of tree line and other vegetation belts affected the equatorial tropics during the Würm maximum. That these were effective down to 2000 m is certain, but the effects upon lower altitudes is less clear. The lowland rain forests, as predicted by Douglas (1969), may have survived reductions in radiation with only minimal changes in structure.

Studies of precipitation variations during the same period, for the rain forest zone involve more uncertainties. Van Zinderen Bakker (1967) has pointed out that each locality has its own rhythm of pluvial and inter-pluvial (drier) conditions, and that these cannot be assumed to have been synchronous throughout the zone. Changes in atmospheric circulation and the position and activity of the Inter Tropical Convergence will have brought different effects to east and west coasts, as well as to interior parts of continental landmasses such as Africa. It is pointed out that the present thermal equator is 5–7°N, and that its position is influenced by the Antarctic ice masses. It seems probable therefore that it will have migrated northwards during thermal maxima of the interglacial periods, when the north Polar ice may have temporarily melted, but the Antarctic ice sheets survived. During the glacial maxima on the other hand, the greater extent of northern ice sheets probably forced the I.T.C. into a more southerly position.

During the retreat southwards of the Inter Tropical Convergence, there must have followed a migration of the northern margins of the African forest belt, possibly to the virtual exclusion of forest conditions from West Africa (Figure 45).

Aubreville (1962) in a review of ecological evidence for the displacement of the rain forest within the inter-tropical zone, has drawn attention to anomalous areas of savanna within the humid tropics of Africa and South America. These include grassy savannas along the West African coast, in Gabon, on the Bateke Plateau and in south central areas of the Congo, and also the extensive savannas of the Rio Branco and Guyana in South America. He attributed these to a time lag in the recovery of the forest, following the drier conditions of the last glacial period. In Africa some of the areas lag behind in the re-establishment of the forest because of locally unfavourable environmental conditions. Aubreville also pointed to the occurrence of savanna woodlands in Angola containing possible remnants of former rain forest. These observations were interpreted in terms of the southerly shift of climatic zones during the glacial periods, and in his reconstruction for Africa dense rain forest conditions survived only in a few 'bastions' north of about 4°S. These were along the West African coast from Ivory Coast to western Ghana, from southern Camerouns into Gabon, and within the marshy interior of the Congo Basin. At this time the main forest belt is supposed to have shifted into Angola, whilst in eastern and central Africa a downward shift and extension of montane forests was apparent (figure 45). Bakker and Levelt (1964) also supported the view that an important climatic change had affected the entire inter-tropical zone of Africa, and also large areas of Surinam, and as will be shown, these authors applied this assumption to questions of landform development.

These changes and their possible causes certainly limit the concept of the stable rain forest ecosystem, but they do not entirely destroy it. However, we cannot assume constancy of conditions over wide areas of present-day rain forest, at least in Africa. It also suggests that on the East African plateaux changes in the structure of vegetation due to reduction in temperatures may have been important at 2000 m and possibly at lower altitudes.

A point of some importance emerges from Walker's (1970) review of vegetation changes in New Guinea, and this is the strong disturbance of natural vegetation during the last 5000 years, and particularly since 2500 B.P., at which date human artefacts are found. The assumption is commonly made that anthropogenic changes to vegetation systems, and therefore to denudation rates, are of such recent origin as to be confined to effects of a superficial or localised kind. These are usually thought of in terms of some degree of sheet

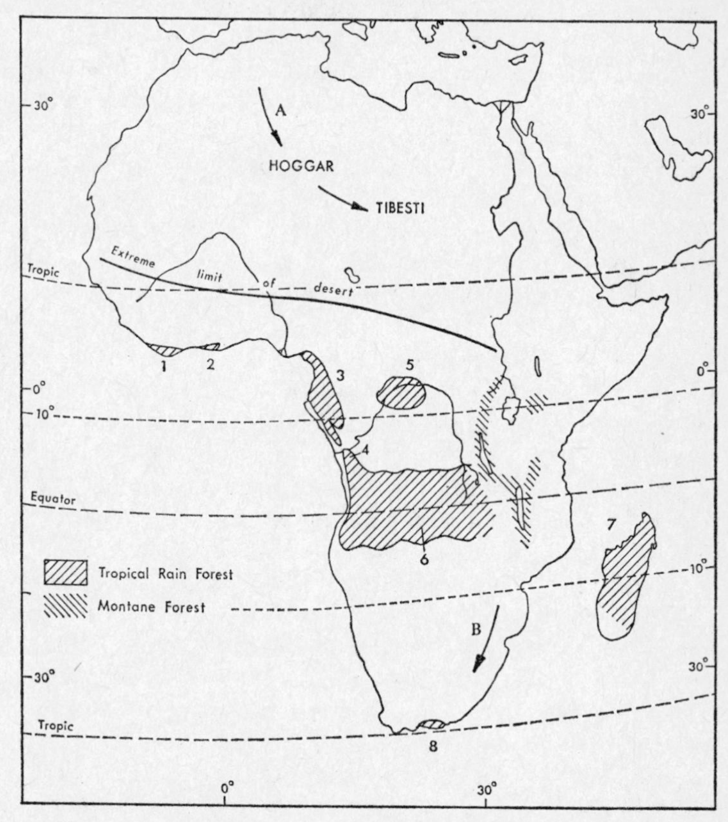

Figure 45 Hypothetical migration of the equatorial zone in Africa during a glacial maximum (after Aubreville 1962).

Forest 'Bastions' are shown:
1. Liberia–Ivory Coast; 2. Ivory Coast–Ghana; 3. Cameroun–Gabon; 4. Angola; 5. Congo (swamp forest); 6. Equatorial forest zone; 7. Madagascar; 8. Tropical forest of Knysna.

Arrows indicate:
A. Migration of Mediterranian flora towards the central Saharan mountains; B. Migration of African montane flora.

erosion, perhaps to expose laterite deposits to induration processes, and to comparable changes in soil structure. Notwithstanding the spectacular effects of local soil erosion, any major change to the landform is usually conceived in terms of periods beyond the apparent influence of man. However, early occupation of the tropical lands by man is well authenticated, and his effects may have been far reaching in terms of vegetation changes, over a period of more than two thousand years. Such a period is probably very conservative, because early human cultures date from a much more remote period than this, but we have little or no knowledge of their impact on the vegetation cover. However, if substantial changes to environment have occurred some thousands of years ago from human intervention, then the effects on rates of waste removal and redistribution across the landscape may have been more fundamental to our understanding of landform development than is generally acknowledged.

Climatic changes at higher latitudes are documented more fully. Mention has already been made of the retreat from present positions of the rain forest : savanna boundary in Africa. Walker (1970) quotes recent work from the Atherton Tableland in Queensland, close to the east coast at latitude 17°S, where from 10 000–7000 B.P. the present broad leaved rain forest was replaced by sclerophyll forest, characteristic of drier environments.

At similar latitudes north of the equator within the continental interior of West Africa this same period was marked by increased rainfall. Grove and Warren (1968, p. 207) conclude that 'the most conspicuous feature of the climatic history of the southern side of the Sahara in late Quaternary times is the pluvial period which was well marked by 10 000 B.P. and came to an end about 7000 B.P. or somewhat later'. In fact features marking climatic change within the savannas of West Africa from about 11°N to at least 20°N can be grouped into four (Grove and Warren 1968, p. 194):

(1) Features probably formed during the middle and early Pleistocene and possibly late Tertiary times.

(2) Forms that originated in major arid periods, the last of which was middle or late Pleistocene.

(3) Features attributable to the last main pluvial, for which dates of between 20 000 and 7000 B.P. are being obtained; and

(4) the more recent events and landforms.

Much of the evidence from this zone comes from sequences of dunes and alluvial deposits. The dune systems extend widely as far as latitude 11°N and locally come farther south than this. But the early Quaternary in West Africa appears to have been marked by extensive alluviation under humid conditions which led to the gullying and dissection of older laterite sheets, many of which date from the middle Tertiary, and to the formation of younger laterites within depressions (Sombroek 1971). Gradually conditions became arid, and north easterly and easterly winds developed the great systems of longitudinal dunes in both the basin of the middle Niger and in the Chad Basin. These dunes, which rise from 10 to 20 m in height, are today reddened by subsequent weathering, and in places may be gullied or breached by permanent stream channels. South of about 16°N latitude, where rainfall exceeds 150 mm, they are now fixed by vegetation growth. The reddening of the dunes undoubtedly occurred during a humid period which led to stream activity as far north as 18°N latitude. Inter-dune depressions became filled by clayey, brown sands, whilst some of the material was redistributed by rivers to form levée systems. These same humid conditions raised the level of Lake Chad by 40 m to form a shoreline feature at 320 m a.s.l., that is, dated to about 10 000 B.P. Subsequently, drier conditions led to the formation of less extensive dunes within the margins of the former 'Mega-Chad' Basin. During the last 7000 years small shifts of climate from greater to lesser humidity have probably occurred to produce comparatively minor features in the landscape. Michel (1967) recognised a short dry phase persisting until 5500 B.P. and Future (1967, quoted by Fölster 1969) a more humid period from 5500–3000 B.P. This may in places have been succeeded by a further dry phase reaching its peak around 2000 B.P. (Michel 1967). As will be shown subsequently, an attempt can be made (Fölster 1969) to relate periods of stability and instability in the landsurfaces of lower latitudes to these events.

Climate and the processes of denudation

It is clear from the foregoing that few landforms or deposits in the tropical landscape have resulted from the uninterrupted operation of a single bioclimatic regime, though perhaps some core areas in the rain forest may have survived significant ecological change for more than 1 million years. It is also clear that present day rates of denudation have been influenced greatly by changes in vegetation cover brought about by man. This anthropogenic influence may have been important over more than two millenia, even if changes during the last two centuries appear to have been most drastic. In fact it can be readily seen that minor climatic fluctuations of the Holocene must have interacted with changes brought about by man in 'natural' ecosystems.

For these reasons morphogenetic zonation must take account of the probable effects of climatic change as this has affected each major bioclimatic zone in the tropics. But because of the problems of inheritance and human land use it is difficult to avoid circular reasoning in any attempt to establish the nature of the denudation system within any particular zone. This problem affects equally the use of contemporary records of stream discharge and load, and also the interpretation of results from morphometric analysis.

Nevertheless, many attempts have been made to characterise the denudation systems of the earth on the basis of present day climatic zones. Nearly all of these hinge upon an evaluation of the vegetation in terms of the protection which it affords the soil and regolith mantle. This depends in turn on two principal features of the plant cover: the proportion of the surface exposed at ground level, and the protection afforded to the ground by the aggregate leaf surface or canopy. In grassland systems these two aspects become indistinguishable. In semi-arid areas, direct exposure of the groundsurface to high intensity rainfall is assumed to lead to rapid rates of surface wash, and the effectiveness of this process is related by several writers (Melton 1957; Schumm 1965) to percentage bare area. As soon as the plant cover becomes continuous Schumm (1965) reasons that rates of surface denudation will decline. The values of rainfall leading to the maximum rate of erosion are given as 325 mm in temperate areas and 600 mm in the tropics, where mean annual temperatures exceed 22°C.

At higher rainfalls the denudation rate is often assumed to be less, though Fournier (1962) argues that it will respond directly to rainfall intensity as expressed in terms of p^2/P, where p is the mean rainfall for the wettest month and P the mean annual rainfall. Fournier also took into account an orographic factor H^2/S, where H is the average height of the catchment and S is the projected area of any catchment over 2000 km². Using climatic data from 650 stations, together with available records of sediment yield from certain rivers, Fournier (1962) derived a number of empirical expressions for calculating what he called 'specific degradation' (ton/km^2/yr^{-1}) over the whole of the African continent. Fournier's map (figure 26, p. 115) is of interest but cannot be accepted without qualifications. It brings out the predicted high erosion intensity within the drier savannas, between 500–750 mm annual rainfall (12°–15° latitude), but in areas of higher rainfall the pattern begs a number of questions. If areas of high relief are excepted, then very high erosion rates are predicted for the Guinea coast from Ghana to Gambia, where rainfall often exceeds 2500 mm but is highly seasonal. This is in contrast to the Congo Basin where similar rainfall amounts occur with much less seasonality. Yet both of these areas lie within the tropical rain forest. On the other hand the wetter savannas of West Africa appear to have low rates of denudation.

Although Fournier suggested that the vegetation factor was accounted for in his formulae, it is not considered specifically. Furthermore, the use of sediment yields from large rivers to predict 'natural' rates of erosion is unjustifiable, in view of the known effects of land use (Douglas 1967). This analysis does however draw attention to the role of vegetation, and particularly to the effectiveness of the tropical rain forest in reducing erosion rates. According to Douglas (1969) and Moss (1969) the rain forest approaches a closed system within which outputs in terms of soil and nutrient losses are very small. On the other hand Young (1972) considers the evergreen rain forest to be the least protective of all the forest communities, and less effective than the humid temperate grassland in reducing surface erosion.

The reasons advanced to account for this apparent anomaly include the absence of a dense herbaceous or field layer, and the rapid decay of leaf litter and dispersal of organic matter within the mineral soil. These features combine to expose the soil surface to leaf drip from the tree canopy, and since many drops are able to fall more than 8 m they reach close to their maximum terminal velocity. During heavy and prolonged storms it can then be argued that rainfall intensity at ground level will soon rise to equal that outside the canopy, even where canopy openings are not important. In addition, water flowing down trunks may cause detachment of soil particles around the base of the tree. Young (1973) has shown that significant surface wash may occur under forest on moderate slopes, but comparative figures for other vegetation communities are lacking.

The case as stated here probably embodies a misconception. It is likely that the rate of surface wash, and perhaps the total denudation rate, will fall as the savanna woodlands merge into semi-deciduous and then evergreen rain forest with increasing rainfall. But within the forested area there are such great variations in rainfall amount and intensity (1500 mm to 3000 mm mean annual rainfall falling in periods varying from 12 to 8 months) that the rate of surface denudation inevitably rises to higher levels within the forest, where precipitation amount and intensity is high, but the protection afforded by the forest remains almost constant (Carson and Kirkby 1972). This accords with Fournier's prediction (figure 26, p. 115) even if his figures cannot be accepted at face value. There can be little doubt too that slope becomes a major factor inducing higher rates of erosion under forest, as elsewhere. Ruxton (1967) showed how leaf litter (which *is* found under rain forest) became rapidly thinner as slope increases in the Papuan rain forest, and Haantjens and Bleeker (1970) demonstrated how rare are mature weathering profiles indicative of slope stability in the same general area.

Given more or less uniform slope conditions, we might therefore predict that the rate of surface erosion will increase to a maximum within the dry savannas and scrub (*c.* 600 mm), then fall gradually as the vegetation cover becomes more protective until it reaches a low value within the less wet forest zone, but increasing again as rainfall attains high values and especially where it remains highly seasonal. However, within these humid areas mass movement may be more important morphologically than unconcentrated surface wash. From such assumptions it is possible also to predict that weathering depths will increase as surface erosion rates decline, and it may be noted that Ruxton and Berry (1961a) predicted such a relationship (see before, p. 3). It might also be suggested that high erosion rates in the savannas will be seen, especially in rapid denudation from interfluves, but a relatively slow rate of channel cutting in streams subject to seasonal flow and high sediment yields. On the other hand, in the high rainfall areas of the forest there may be relatively strong channel cutting from consistently high runoff, but rather localised denudation on slopes. These highly qualitative remarks must be treated with caution, but point to possible differences in denudation systems between bioclimatic zones, and suggest that as these zones have oscillated latitudinally, so the rate of erosion will have fluctuated and given rise to morphological changes, some of which are considered below.

Geomorphic effects of climatic change

Shifts of the climatic belts by 5° latitude or more, would have brought much higher erosion rates to areas such as southern Nigeria and Ghana, and the northern part of the Congo Basin, during the Pleistocene. This supports the possibility that the stripped etchplains of southern Nigeria (Thomas 1969) may have formed under conditions of more severe erosion than obtain at the present time. Conversely, the surviving remnants of ancient regoliths in the Sudan zone, recorded long ago by Falconer (1911), may be evidence for the former dominance of humid tropical conditions in these latitudes throughout much of the Tertiary period (Millot 1970).

These observations and arguments also lend some support to the contentions of Bakker and Levelt (1964) that stripping to form or maintain bare rock domes in the tropics requires a seasonal climate. They followed two lines of reasoning: first, that excavations of inselberg footslopes in Surinam showed a thin weathering profile developing over buried, sub-aerial weathering forms such as *Koçiolki* (weathering pits). This suggested to Bakker and Levelt that the inselbergs were relics of a drier period of savanna climate and that the flatter granite slopes were becoming colonised by rain forest, following a recent change to more humid conditions. Secondly these authors found illitic clays (20 per cent illite) and the heavy mineral epidote in lateritic formations on certain granite summits. Illite does not readily form under humid tropical conditions, and epidote is unstable where the soil temperature is high but the pH is low. They were thought therefore to indicate a quite dry, savanna climate.

On the other hand ecological studies on inselbergs have been inconclusive. Hambler (1964) found evidence of progressive colonisation of slopes on a domed inselberg within the savanna of southern Nigeria, where mats of the sedge *Trilepis pilosa* appeared to be giving way to the hydrophile grass *Andropogon linearis,* and later to savanna communities of increasing complexity. On the other hand Richards (1957) working near Idanre (figure 30, p. 169) where perhaps some of the most striking domes in the world occur well within the present rain forest boundary, found that the pioneer communities of *Trilepis* were separated by sharp discontinuities from forest communities, nourished by soil in gullies and clefts between the domes. There was also some evidence of periodic removal of *Trilepis* mats to leave bleached patches on the bare rock.

Freise (1938; quoted by Bakker and Levelt 1964, p. 42 footnote) thought that, within the rain forest, a cycle of development and decay of vegetation would cover and uncover the inselberg domes. The stages were:

(1) bare granite and gneiss domes and ridges;
(2) accumulation of humus from lichens and algae and the development of savanna;
(3) immigration of trees;
(4) development of tropical rain forest and strong weathering;
(5) degradation of the forest with crust formation and strong soil erosion;
(6) deterioration of the forest to produce savanna communities;
(7) domination of grassy vegetation and destruction of soil profiles;
(8) emergence of inselberg domes and ridges (return to 1).

There seems to be little evidence to support this as monoclimatic concept, but Bakker and Levelt (1964) thought that, with climatic change such a cycle could come about. The nature of the climatic oscillations considered in this chapter could certainly be construed to support such an hypothesis.

But there remain some further questions, for although extensive stripping to form etchsurfaces in the humid tropics may be unlikely, localised removal from resistant kernels of rock can be seen to occur as a result of landslides (Plate VII, and figure 35, p. 193) after Hurault 1967). Some of the scars may be rapidly recolonised by vegetation and soil, but this is by no means certain to affect all or even most of them. There remains, therefore, a possibility of dome formation within the forest. Furthermore, domes formed during earlier and less humid climates may survive indefinitely within the rain forest; their arid rock surfaces, able only to support xerophytic plant communities, and the rapid runoff from the rock surfaces periodically removing colonies of more sensitive vegetation. There are also some rock pavements appearing within the forest of southern Nigeria, and these can hardly date from even the most recent sub-humid phase for they are accordant with the present landsurface.

It has already been shown (chapter 5, chapter 9, and p. 252) that widespread slope deposits occur on tropical plains, and that these do in some cases appear to exhibit a clear stratigraphy. They have been related to periods of landsurface stability and instability in both Australia (Butler 1967) and Africa (Fölster 1969). Similar sequences have been identified in high latitudes and Starkel (1966) for instance has related these to late Pleistocene and Holocene climatic changes.

Opinions differ however concerning the significance of the 'K-Cycles' recognised for parts of Australia. Van Dijk *et al.* (1968) found that stratigraphic principles could not be applied in broken country around Canberra, but that the age of the deposits might be related to position in the landscape: the oldest deposits being found on upland surfaces with progressively younger materials encountered towards lower slopes.

In the African tropics, however, Fölster (1969) has attempted a chronological framework of slope erosion and deposition, but his chronology is based on analogy with events in the Chad Basin. Apart from possible earlier remnants on sedimentary outliers near Ilorin, Fölster recognised the following sequence in south-western Nigeria:

C —(stable) crusts, representing lateritisation during a relatively remote period, and forming scattered interfluve remnants;
B —(unstable) removal of C- soil cover;
—(stable) crusts, representing a widespread development of a ferruginised layer, found almost exclusively in drainage depressions on pediments cut into the old C- profiles;
A1 —(unstable) partial removal of B- soil mantle and deposition of pediment gravel and hillwash on new pediments;
—(stable) weathering and local cementation of gravel and hillwash;
A2 —(unstable) removal of most of A1 hillwash;
—(stable) pedogenic alteration of redeposited hillwash, with mottling and clay eluviation;
A3 —(unstable) with shallow, localised erosion of A2 hillwash;
—(stable) present soil formation.

In each unstable or erosional phase, backwearing of low scarps in indurated material tended to sort coarse gravel fragments, derived from the crusts from finer material which was laid down on top as hillwash. Fölster obtained a single C^{14} date of 2360 ± 120 B.P. for the most recent soil layer near Ife. This suggests that at least some of the older soils have a considerable age, and have not resulted from very recent anthropogenic disturbance. It also implies that earlier deposits are likely to have resulted from climatic changes during the Pleistocene and Holocene. The following tentative relationship with events in the Sudan zone was offered:

TABLE 16

SUGGESTED CORRELATIONS OF MORPHOGENETIC EVENTS WITH CLIMATIC CHANGES IN S. W. NIGERIA

Geomorphic activity	*Climate*	*Date* B.P.
A3—Soil fixed by recent vegetation	present	
—Erosion: parabolic dunes in Senegal delta	arid	after 2500
A2—Soil development: slightly higher lake levels	humid	
—Erosion: re-activation of dunes, recession of lakes	arid	5500–3000
A1—Weathering: late glacial reddening of dunes, high lake levels	humid	12 000–8000
—Erosion: Würm dune fields formed	arid	*c.* 30 000
B—Crust formation	humid	
—Erosion	semi-arid	
C—Crust formation (from Riss : Würm interglacial)	humid	before 60 000

(Mainly after Fölster 1969)

Burke and Durotoye (1971) present a similar but less detailed sequence of events based on evidence from sections within the same general area. They recognise a common sequence from weathered rock through a ferruginised or Older Pediment Gravel, above which is found a Younger Pediment Gravel which remains uncemented, and is covered by the sandy loam deposits, discussed above as a product of hillwash, termite and earthworm activity. Correlations are suggested for these deposits on the basis that major climatic fluctuations would be required. The Younger Pediment Gravel was therefore related to the ancient dune fields of the Sudan (20 000+ B.P.), and thus appears to correspond with Fölster's A1–unstable phase. The ferruginised Older Gravels were placed tentatively at circa 60 000 years to correspond with the Würm glacial maximum.

Such a chronology is only a first approximation, but it emphasises the probability of increased erosional and depositional activity within the moist savannas during the drier phases of the last glacial. It is also arguable that similar and perhaps more severe fluctuations occurred during earlier glacial/interglacial oscillations. It also seems likely that accelerated removal affected the forest zone at this time, and the presence of widespread colluvium is confirmed by field observation and by Moss (1965; and figure 17, p. 78).

On and around the Jos Plateau in Nigeria (latitude 11°N) there are also striking accumulations of material. This is an area of high erosion intensity at the present time (Fournier 1962; and figure 26, p. 115), and mention has already been made of extensive fan deposits at the foot of the escarpment zone (figure 31, p. 171). In 1949 Bond described some of the alluvial sequences in the tin deposits of the Plateau. Unlike southern Nigeria, this is a region of high relief, and also one which has experienced important vulcanism, starting probably during the Tertiary Period and continuing into the Pleistocene. Nevertheless Bond was able to point to three periods of repeated events: (a) erosion; (b) deposition of gravels with tin; (c) deposition of sands.

This sequence occurs in all the stream valleys, and appears to record decreasing stream energy which Bond (1949) related to periods of decreasing rainfall. The lowest sequence at Nok shows lateritisation of the sediments, and each sequence has in places been subject to erosion. Bond also attempted to relate these events to those of East Africa by means of type implements. However, it is clear from recent work in the Kenyan Rift Valley that a climatic rhythm cannot entirely explain the sedimentary sequences there (McCall *et al.* 1967). Breaks in the succession are in fact nearly everywhere due to episodes of tectonic activity, and although climatic changes have also been of great importance, to attempt any résumé of the evidence for the East African area is beyond the scope of this study (see Bishop and Clark 1967).

Some implications for landform development in the tropics

It is first necessary to stress that within the latitudes that remained immune from eolian activity during the Pleistocene, the changes of temperature and rainfall that took place could only result in variations in the balance between weathering and soil development on the one hand, and surface removal and soil truncation on the other. No fundamental alteration to the landforming processes occurred. Secondly, it must be acknowledged that the shifts of climatic and vegetation zones affected most of the intertropical zone. In Africa rain forest conditions probably persisted in only a few areas of no great extent. But the migration of climate would also have brought forest to new areas, where it may have survived for some time during intervening dry periods. From this it is clear that periods of increased erosion would not be synchronous throughout the tropical zone during the climatic oscillations. Reference to Aubreville's map (1962; and figure 45) suggests that the direction of change in Africa during a single swing of the climatic belts could be:

TABLE 17

GENERALISED GEOMORPHIC RESULTS OF CLIMATIC CHANGES FOR TROPICAL AFRICA

Direction of Change	*Geomorphic Results*
northern savannas→desert	eolian conditions: dune development
wetter savannas→more open grassy vegetation	accelerated erosion; stripping
outer forests→wetter savanna	accelerated erosion; stripping
central forest belt→some areas stable	steady state
wetter savannas→forest (of southern hemisphere)	decreased erosion, soil development

As the climatic swing reversed, so the original conditions would return. But we know from areas such as the Chad Basin that climates have been both more arid and more humid in the past, so that the return swing leads to increased humidity of climate in the northern hemisphere zones shown. Such rhythms have probably been very widespread and may possibly be used to define morphogenetic zones on the basis of a repeated pattern of change and disturbance in ecological conditions:

TABLE 18

POSSIBLE MORPHOGENETIC ZONES DEFINED ACCORDING TO THE PATTERNS OF CLIMATIC CHANGE

Decreased humidity	*Zonal Climate*	*Increased Humidity*
Stable (Desert)	Desert (core)	Stable (Desert)
Stable (Desert)	Desert (margins)	Open Savanna
Desert	Open Savanna	Savanna Woodland
Open Savanna	Savanna Woodland	Rain forest (dry)
Savanna Woodland	Rain forest (margins)	Rain forest (wet)
Stable (Rain forest)	Rain forest (core)	Stable (Rain forest)

It is perhaps relevant to indicate that, according to Fournier's analysis of erosion rates, drier conditions would initiate an acceleration of erosion on slopes within both the forests and the wetter savannas. But in the Sudan zone of drier savannas there would be a decrease in fluvial activity of all kinds, with a corresponding increase of wind action. Accelerated erosion would therefore be likely to correspond with the wetter periods in this zone. However it is also within these periods that soil development and weathering would be favoured by increased humidity of climate. This situation possibly complicates the K-Cycle concept.

Of the earlier and perhaps more major changes we know less, though the formation, removal and re-forming of laterite sheets may be examples of such events. It has been suggested that the deep weathering mantles of the tropics and sub-tropics may date from the Tertiary or even the Mesozoic Eras (e.g. Ollier 1969), whereas some of the indurated slope deposits appear to be late Pleistocene or even Holocene in age. The conclusion is unavoidable that very deep weathering profiles require very long periods of geological time to develop, whilst rapid removal of material may effectively strip large areas of the basal surface of weathering in a comparatively short period.

Areas of the tropics which experienced humid conditions throughout the greater part of the Tertiary Era, but which were exposed to high erosion intensity during much of the Quaternary have probably lost a high proportion of their former regolith cover. Some of this material is probably still present on the surface as transported accumulations. On the other hand the inner zone of the rain forest may have retained much of its regolith cover, and perhaps more important, is capable through contemporary weathering of maintaining a weathered mantle on steep slopes.

The emergence of the types of etchplain enumerated earlier in this study (figure 41, p. 236) is probably therefore dependent upon many of the changes in climate and balance of geomorphic process recorded here. It is important also to stress that the *disturbance* of conditions has much to do with sediment yield and morphogenetic change, and periods of change are more likely to mark periods of most effective surface erosion than are any stable conditions, whether savanna or forest, that have persisted in between.

It is difficult to generalise concerning the landscape patterns resulting from periodic fluctuations in the rate of surface denudation, although one aspect of these events is embodied in the simple model of progressive stripping to form the varieties of etchplain shown in figure 41. One feature of the erosional effects appears to be the thinning of ancient regoliths on interfluves, or the gradual removal of weathered material from flanking slopes as duricrust breakaways retreat within the waste mantle. The corollary of these events is a corresponding pattern of deposition which has also received comment here.

It is not always possible to demonstrate clearly the relationship between these two patterns unless a correlation can be established between the *in situ* and transported materials. Thus, it has been suggested that fan accumulation below the scarps of the Jos Plateau (figure 31, p. 171) can be related to the advanced stripping of the upper escarpment to form tor and dome landscapes, but the evidence for this is largely circumstantial. On the other hand Churchward (1969) was able to demonstrate that transported soils within the valley systems of Western Australia can commonly be related back to zones within the relict weathering profiles. But particularly in this latter case the effects of uplift may dominate those of climatic change (Mulcahy *et al.* 1972). This problem is met with in western Nigeria where the etchplain model (figure 41) was developed, for this is an area of Tertiary uplift and warping which must have led to increased gradients in south flowing streams.

It should be possible, however, to distinguish the effects of base level changes from those of widespread landsurface instability (Thomas 1965b). In the first case redistribution of mantle deposits will radiate from the drainage network, and the effects on interfluves will depend on depth of incision and drainage density. On the other hand, where open savanna conditions are involved, climatically induced instability is likely to lead to widespread redistribution of deposits across gentle slopes on old landsurfaces and interfluves between drainage lines, and to the removal into the stream channels of large quantities of progressively coarser debris.

On valley sides such events may interact to produce confusing patterns, and in many cases, including those cited here, both forms of activity in the landscape have occurred either concurrently or in succession. Even where diastrophism cannot be indicated, base level changes are likely to have occurred because of world sea-level changes during the Pleistocene, so that climatic change must inevitably bring to most areas changing conditions of both landsurface stability and stream activity. The latter will result both from changed runoff relationships and from base-level changes.

11 Weathering and Planation in Extra-Tropical Areas

Reference has already been made to the recognition of global weathering patterns based upon climatic controls (Bakker 1967; Pedro 1968; Strakhov 1967), and to some of the deductive schemes for the definition of morphogenetic regions (Peltier 1950; Büdel 1957, 1970). Although some of these ideas have been shown to rest upon slender field evidence (Stoddart 1969), enough of substance remains to support the recognition of specific relict landforms and weathering mantles now found outside their regions of characteristic development. Most of the features which have attracted attention relate to the occurrence of deep weathering (Fitzpatrick 1963; Demek 1964); to the nature of weathering products (Bakker 1960, 1967; Bakker and Levelt 1964), or to the recognition of relict land-forms such as tropical karst residuals (Pokorny 1963), and pediments (Büdel 1957; Gellert 1970). However, as Dury (1971) has recently pointed out, the broader implications of much of this evidence have been neglected by all but a few English speaking geomorphologists, and the findings of many European writers have received scant attention.

There are in fact at least two distinct groups of problems. The first is the doubt which must still surround the climatic significance of deep weathering which as Dury (1971) points out is difficult to define. The work of the late Professor Bakker and his colleagues at Amsterdam has gone far towards a more adequate appreciation of the range of weathering types in Europe and their correspondence with tropical and extra-tropical regoliths elsewhere. But deep dis-aggregation of granitic rocks for instance may not be restricted to the warmer climates. It seems likely that the deep sandy regoliths (arenisation) described by Bakker (1967) have a very wide occurrence in mid-latitude lands, and that their development may, in some cases at least, be related to original weaknesses in the rock fabric (Eggler *et al.* 1969). This phenomenon may not therefore have any well defined climatic significance. But whether or not most deep weathering has occurred under sub-tropical or tropical conditions is less important than the recognition of its widespread occurrence as a phenomenon in most extra-glacial areas.

The second group of questions surrounds the importance of relict weathering mantles and palaeoclimates to the understanding of present-day relief, both in glacial and in extra-glacial areas beyond the tropical zone. Some aspects of these problems are dealt with in this section.

Records of deep weathering

A most striking early opinion given by Chalmers in 1898, on the basis of observations in eastern Canada, is quoted by Feininger (1971) in an interesting study of these problems. Chalmers concluded his study (1898, p. 280): 'The occurrence of such extensive sheets of decomposed sedentary rock in the region under consideration, much denuded as it may seem to be, points to the former existence of a universal mantle of this material overspreading the country everywhere in Tertiary and preceding ages.' Such a claim is not contradicted by more recent work, and if accepted, demands consideration alongside Falconer's similar observations made under tropical conditions in Nigeria (Falconer 1911). Such concepts are in a sense the starting point for the study of tropical landforms, and we must therefore consider carefully how far the approach to tropical conditions may be extended into higher latitudes. Feininger (1971) considers that a lateritic blanket with an average depth of around 30 m may have covered much of the northern continents prior to glaciation, while Dury (1971) points out that relict crusts have been observed from 43°S in Tasmania to at least 50°N in Europe and America.

The depth of surviving mantles is highly variable. Kaye (1967) notes kaolinitic decomposition of phyllites near Boston, Massachusetts, to a depth of 91 m. Demek (1964) has noted a similar depth of weathering in Bohemia, but generally surviving regoliths are much thinner than this. Most of the records referred to by Bakker (1960, 1967) indicate depths of 6–10 m from Morvan in central France and the Harz Mountains in Germany. Similar depths are found widely in central Europe and even at scattered localities in Scotland (Fitzpatrick 1963). Demek (1964), however, quoted many figures for kaolinitic weathering in Czechoslovakia in excess of 20 m, and he considered that widespread removal of a former tropical regolith following Neogene uplift has left only the basal parts preserved in irregular masses usually of 10–20 m in depth. In the Laramie Ranges of Wyoming, Eggler *et al.* (1969) found up to 60 m of rotted granite at over 2300 m above sea level (41°N). Ollier (1965) also found great depths of weathering at high altitudes in the Snowy Mountains of New South Wales.

The possibility that any large proportion of this weathering can be recent must be limited to zones of slight alteration in shattered or weakened granites, where the formation of a deep, sandy gruss might possibly have occurred quite rapidly (perhaps within a few thousand years?). But knowledge of weathering rates suggests that most weathering proceeds slowly. Feininger (1971) quotes interglacial weathering of drift deposits of from 3–4 m during perhaps 100 000 years. In the tropics Leneuf and Aubert (1960) suggested 22 000–77 000 years for 1 m of gneiss to become thoroughly ferrallitised. Such figures support the notion that most of the relict profiles are pre-Pleistocene in age, especially those in which a high clay content is found.

Bakker (1967) made a fundamental distinction between the sandy weathering (arenisation) and clayey weathering over granites in central and western Europe. In the former only 2–8 per cent of clay is found and many feldspars remain intact. The clay minerals are usually illite (40–50 per cent) and kaolinite (40 per cent). Bakker considered that this regolith might well date from the early Quaternary, having formed under a western Mediterranean climate, analogous with Atlantic Portugal at the present day. However, the clayey weathering, possessing 15–30 per cent of clay, in which kaolinite is frequently dominant, was thought to correspond with a wetter and warmer climate found today as far as 20°N latitude along eastern seaboards of the continental landmasses. This type of weathering is recorded from the Ardennes, where kaolinite forms 50 per cent of the clay fraction, and in Thuringia, the Harz and Spessart Mountains, where clays containing up to 80 per cent kaolinite have been found. In all cases the absence of gibbsite and the presence of perhaps 20 per cent illite suggest a climate less hot and humid than the tropical forest, or monsoon types, in which gibbsite commonly forms up to 30 per cent of the clay. Bakker pointed out that recent superficial clay formations in the brown forest soils of Europe are dominated by illite. These observations are corroborated by those of Pokorny (1963), who found that fissure clays around limestone residuals on the Cracow Upland of southern Poland contain up to 90 per cent kaolinite, whilst superficial clays from intervening depressions indicated only 15 per cent kaolinite, and 60 per cent illite.

The possible dating of these clays depends on independent geological evidence. In Poland some of the kaolinitic clays are found in pockets beneath Tortonian deposits and have been considered Palaeogene in age. On general reasoning, Bakker and others considered many of these deposits to date from the Eocene, though some are almost certainly more recent. Dury (1971) considers some of the independent evidence, and points to the existence of the Red Beds in the Antrim Basalt sequence, and the Indo–Malay fauna of the London Clay, both of which date from the Eocene. He points also to the Arkansas bauxites which may have resulted from early Eocene weathering, and to the similar deposits of southern France which may be even older. On the collective evidence Dury (1971) considers that the early Eocene may have been a period of maximum warmth in Europe with mean annual temperatures as high as 20°C, but cooling during the Oligocene, and after some recovery during the Miocene deteriorating progressively until the end of the Pliocene by which time they may have been as low as 10°C. Dury thinks that temperatures as high as 18°C may have reached to 50°N and were accompanied by high humidity.

If some of the dating remains uncertain, the weight of evidence for extensive pre-Pleistocene deep weathering is overwhelming, and its influence on the course of landform development both at the time and subsequently has been important.

The fundamental contribution which weathering processes make to landform development in the tropics may clearly be extended with shifts, merely of emphasis, into the sub-tropical areas; for instance of south-western and south-eastern Australia, from which areas many examples in this book have been drawn. The only major discontinuity in developmental history and process comes, in fact, at the limits of northern hemisphere glaciation (White 1971). It is therefore perhaps appropriate to consider extra-glacial and glaciated areas separately.

Weathering and landform development in temperate extra-glacial areas

The contribution of weathering to landform development in these areas has long been recognised, particularly in the history of granite tors and similar formations. The early accounts will not be reviewed here, but they show in many cases that attention was directed to this factor more than 100 years ago. Notwithstanding the controversies surrounding tor development on Dartmoor and the possibilities of both hydrothermal alteration in the granites, and periglacial altiplanation resulting in rock stacks (Palmer and Neilson 1962), Linton's (1955) study remains relevant to accounts of this and similar topographies. The exhumation of tors from deeply weathered granite has been claimed also from central Europe (Demek 1964), from western North America (Cunningham 1969), and from many other localities. In fact an exhumation hypothesis for schist tors in Otago was proposed by Cotton as early as 1917, and was supported by a more detailed study made by Ward (1951). The deep weathering was thought by these writers to date from the Tertiary, and red clays are certainly found in Tertiary sediments in this part of New Zealand. There is evidence too of an ancient silcrete formed in association with the tors in Otago, and Dury (1971) has called attention to the possible interpretation of English 'sarsen' stones as silcretes. Deep granite weathering and tor topography are of course widespread throughout eastern Australia, where again some of the tor topography has been dated as Tertiary (Browne 1964).

Residual hills of other kinds have also been attributed to tropical or sub-tropical weathering processes, especially in Europe. Pokorny (1963) has described residual limestone mogotes from southern Poland and, as quoted above, showed that residual Tertiary clays had formed in association with these hills which resemble tropical karstic mounts.

The widespread occurrence of inselberg forms in central Europe has recently been reviewed by Gellert (1970). Most of these surmount surfaces known to be of Tertiary age, and have been described by many European geomorphologists from areas such as Saxony, Silesia and the Sudetes. Jessen (1938), Büdel (1957) and others have attributed such surfaces and their residual hills primarily to tropical or sub-tropical weathering and planation, active during the middle Tertiary period. Bakker and Levelt (1964) have reviewed much of the evidence for Tertiary climates and climatic changes in central Europe and concluded that not only were tropical conditions prevalent during this period, but also that important variations of climate occurred with a periodicity of from two to six million years. These variations were associated mainly with humid-arid oscillations, semi-arid conditions prevailing in the lower middle Oligocene and Miocene. Such oscillations were thought by Bakker and Levelt (1964) to have been capable of developing both the deeply weathered etchplains and the pediments of central Europe. They also emphasised the importance of the climatic variations in bringing about landsurface instability. It may be doubted whether the climate throughout this long period was truly tropical. The warmer climates of the Oligocene may have resembled those of the Mediterranean, but with higher temperatures and greater rainfall. Such a winter rain regime might have different effects than those apparent in the tropical savannas. Gellert (1970) speculates that such a climate might account for the gentler slopes of residual hills in central Europe, but such details remain unclear.

Büdel (1957) goes as far as to claim that plain formation is largely confined to the seasonal tropics, where morphogenetic conditions favouring plain formation have persisted throughout a long geomorphic history. These tropical plains are characterised by extensive deep weathering and are bordered by rocky wash pediments around inselbergs and escarpments leading towards higher surfaces. Such pediments forming *Rumpftreppen* are recognised in central Europe by Büdel (1957) who claims that they are not remnants of former, extensive rock plains, but rim pediments bordering weathered surfaces now largely destroyed. The stripping of the basal surface of weathering has, according to Büdel, exposed the *Grundhockerrelief* on many European upland landsurfaces, as in Swabia and Franconia. Such an hypothesis implies the development and survival of etchplains or etchsurfaces widely in Europe and presumably in other high latitude areas. Büdel emphasises the importance of chalk and limestone residual hills as indirect evidence for these events.

It is clear from all these accounts that evidence for tropical weathering and planation may be dated mainly to the middle Tertiary, while the stripping and dissection may be associated respectively with periodic aridity in the Tertiary and a progressive cooling of climate during the Pliocene and early Pleistocene. The great length of this period of landform development appears to have left a legacy of etchplains, pediments and inselbergs as forms inherited by more recent temperate and periglacial denudation systems which have greatly modified this Tertiary relief, leaving only fragmentary evidence of its former widespread existence.

Weathering and landform development in formerly glaciated areas

The influence of preglacial deep weathering upon the course of glaciation has been only sporadically mentioned in the literature. Bakker (1965) drew attention to the probable effects of differential deep weathering on the depth of glacial erosion in valleys, attributing some glacial steps to this cause. More recently Feininger (1971) has made a direct analogy between the morphology of tropical deep weathering and the character of basined landscapes and till from glaciated areas. Particularly over areas of low relief, Feininger (1971) argues that continental ice sheets did little but remove a deep regolith from the irregular basal surface of weathering. In this way basins and troughs in the ancient weathering front have been exposed, or later filled with water, to form the familiar lake studded landscapes of the glaciated shields and uplands.

Such an hypothesis may appear tenuous when confronted by the weight of evidence for glacial erosion, and it is directly challenged in another recent and wide ranging analysis by White (1972). This author, whose interesting early papers on bornhardts (1944, 1945) have already been recalled, considers that the absence of these residual hills from the glaciated areas of the northern shields can only be explained in terms of the efficiency and pervasiveness of glacial erosion, close to the centre of the continental ice sheets. He develops a zonal model to elucidate the major morphological features of the formerly glaciated areas, and if this is accepted, then the contribution of pre-glacial weathering to these areas must be small, having a transient effect only at the onset of early glaciation. This study cannot attempt to resolve this conflict of views: they possibly represent the extremities of current argument. One can only comment that intensity of glacial scour was very varied, and that although the argument advanced by Feininger (1971) should not be pressed too far, there is evidence to support its application in some areas.

Thus, in a study of Finnish Lapland, Kaitanen (1969) thought that certain anomalous features of till composition corresponded with characteristics of weathering crusts and that tors and other landforms on the upland surface of northern Lapland may have been developed by a combination of exhumation and pediplanation. In a comparable study of the Paleic Surface of Norway, Gjessing (1967) refers to a study by Reusch in 1903 (!) in which many hills and depressions were attributed to former differential deep weathering. Gjessing also reviews the possibility that two of the major form elements of the old surface: residual hills, and basined depressions, may have been formed in preglacial times by a combination of deep weathering and pediplanation. This study makes specific reference to the literature of tropical geomorphology and implies that such considerations may have much broader application to the study of parts of Sweden, Scotland and elsewhere.

Although old weathered mantles are only occasionally preserved at these high latitudes (Fitzpatrick 1963; Kaitanen 1969) there can be little doubt that preglacial deep weathering has been influential in both the modelling of ancient surfaces and in guiding the course of glacial and periglacial denudation. It is of some interest to compare the attempt by Mabbutt (1965) to trace the ancient weathered landsurface across tropical, central Australia, with the work of Gjessing (1967) in which he traces the extent of an ancient 'Paleic' surface across Norway. This approach, whereby an attempt is made to recognise a *complex* of landforms and deposits having a common origin and development throughout comparatively stable areas, reveals much about the early development of major forms and their persistence in the contemporary landscape. By looking at the nature of the inheritance and the extent of inherited forms, many of the objections to the mapping of supposed planation surfaces are overcome and a convergence between the study of process in landform development and denudation chronology becomes possible. In this convergence the contribution of tropical studies to an understanding of high latitude relief is fundamental. What is not always realised is that progress along these lines was made more than 60 years ago. But with the great controversies on slope development, insistence upon a strict uniformitarianism, and later emphasis upon zonal morphogenetic regions, our appreciation of detailed form character and recognition of the oscillatory nature of morphogenetic systems has long been delayed.

12 Some Problems of Landform Study in the Tropics

Many urgent problems encountered in geomorphological work within the tropics are of a practical nature, and are concerned with the availability of base maps and aerial photographs, accessibility or observation within forested terrain; added to which are the problems of sub-surface exploration, where weathering depths may exceed 50 metres. As a result, we lack sufficient field data (Tricart 1972) for many analytical procedures. Furthermore, as Clayton (1970) has emphasised, one's perception of the landform depends often on the conviction which comes from aquaintance with the field situation. Individual perception of the range of conditions within the tropics is therefore severely limited, yet enough can be understood to recognise fundamental and theoretical problems, some of which are briefly outlined in this postscript.

Some features of weathered landsurfaces

Much of this study has been concerned with deeply weathered landsurfaces which have experienced prolonged sub-aerial histories. These terrains present a number of problems of interpretation. In the first place, there can be little doubt that if characteristic landform assemblages occur in the tropics, they will be found in those areas where weathering under tropical conditions has continued over a long period. But few comparisons are available from higher latitudes that would allow an assessment of such landscapes in a zonal morphogenetic context. Nevertheless, it appears conceptually sound to regard these landscapes as the products of a limited range of environmental conditions. Available evidence from other areas does not allow the conclusion to be drawn that they are a function only of time.

Some arguments concerning the development of such landscapes are affected by lack of knowledge concerning the 'initial' conditions from which the present pattern of landforms has developed. Concepts of deep weathering and stripping lead to the reconstruction of former, weathered landsurfaces from which the present pattern of weathered plains, and rock landforms have been exhumed. But this is an assumption which may not be supportable. Irregular weathering patterns are known to occur beneath sedimentary rocks in Nigeria (Jeje 1972), and Savigear (1960) has suggested that some inselberg landscapes may have been revealed as the sedimentary cover was stripped from the Basement Complex in southern Nigeria. Thus some patterns of deep weathering and rocky inselbergs may have developed not from some former surface of planation, but from similar patterns at a higher level, and derived by exhumation as previously described. Against this view may be placed the concept of a supra-continental 'Gondwana Surface' (King 1962) which may have been the product of prolonged denudation during the early Mesozoic, and from which the present complex of forms could have been developed. Such a surface might indeed have been widely and deeply weathered across many rock formations, but there is little surviving evidence for this. On the other hand extensive weathered plains, transgressing important petrographic boundaries, do exist today, and others have been tentatively reconstructed (Mabbutt 1965).

Because of the doubts surrounding such speculation, it is more profitable to concentrate upon the detail of existing forms and deposits and the evidence for their immediate origins. Furthermore, because the concept of pediplanation concerns a periodicity in landscape development extending over 10^7 years, to focus attention on this question leads to the neglect of such details. The surfaces extending as the escarpments retreat must logically vary in age from the very recent to the most ancient, and to describe them simply as pediplains ignores their later development. Many such planation surfaces have co-existed at different levels in the landscape throughout much of Tertiary and Quaternary time, and except where climate has been modified by altitude, they may exhibit similar patterns of weathering, and carry similar deposits of laterite or transported colluvium. In other words the time scale of planation is so extended and the formative events so infrequent, that they account only for the gross form of the landmasses. Any detailed understanding of their landforms and deposits must therefore be derived from a study of their continuing development, and often of quite recent events.

It is instructive to compare the span of 'cyclic' time as described by Schumm and Lichty (1965) or Ahnert (1970), with the short 'K-Cycles' of alternating soil development and instability (Butler 1959). The former take place within 10^6–10^8 years and over areas of 10^2–10^4 km^2, while the latter recur within periods of from 10^1–10^3 years and affect areas of less than 10^{-1} km^2.

Thresholds of erosion and accumulation

The coexistence of stripped and weathered surfaces over similar rocks, makes it clear that rates of surface denudation have varied from below rates of weathering penetration to values ensuring the extensive removal of regolith. But our knowledge of these rates is primitive, and our understanding of the thresholds which control alternating periods of stability and instability in the landscape is equally limited. This is a serious gap in our knowledge for it is a major part of the reasoning within this study, that concepts of biostatic equilibrium and of rhexistatic disturbance, are necessary to account for observed patterns of landforms and deposits (Millot 1970).

Attempts to describe rates of weathering have been frustrated by the slowness of decay and problems of defining the degree of alteration achieved in a given time. Haantjens and Bleeker (1970) to some extent corroborate the order of magnitude (1 m in 22 000–77 000 years) for ferrallitisation offered by Leneuf and Aubert (1960). If their figures establish the likely limits, then surface removal must have been very slow over a very long period to give rise to observed depths of duricrusted weathering profiles. On the other hand, deep sandy gruss may form quite rapidly in certain granites, and it retains a considerable degree of stability even in dissected terrain with slopes exceeding 20°. It is arguable that this deep disaggregation took place during dissection, and partly as a consequence of increased groundwater movement resulting from relief development. Other profiles may have deepened during dissection of duricrusted landsurfaces.

Increased mobility of regolith generally accompanies the advance of weathering, so that kaolinised, clay residues are likely only to arise under conditions of low relief (Haantjens and Bleeker 1970) and a slow pace of surface erosion. Slight oscillations of climate and vegetation cover appear able to induce instability on such regoliths, leading to local redistribution of the upper zones of the profile. Quite different thresholds exist for the highly permeable gruss, having less than 8 per cent of clay. Considerations such as these make more difficult the task of interpreting the patterns of outcrop, regolith and deposits within the landscape.

Inheritance and landscape dynamics

These factors also contribute to the occurrence of forms and materials in the landscape which are inherited from an earlier morphogenesis (Ruxton 1968). The reasons for inheritance are varied, but indicate that many of the components of the landform do not respond rapidly to changes in the controlling denudation system. Recognition of relict regoliths is but one aspect of this topic, and according to some interpretations, ancient regoliths have been progressively removed from the tropical and sub-tropical shields over a period of 10^8 years (Ollier 1965, 1969).

Other forms of inheritance are much more recent. Drainage density for instance may in many areas represent a channel network adjusted to an earlier, more humid period. Such periods occurred during the Pleistocene, and as recently as 10^4 years ago. Other aspects of the channel system may, on the other hand, be a response to forest clearance during the last 10^2, or at most 10^3 years. In a similar way landslides may reflect events of the Pleistocene; the incidence of earthquakes at 10^3 year intervals; the effects of storms having a 10^1 or 10^2 year recurrence, or the more or less immediate results of human interference with natural slopes or plant cover.

The importance of exceptional events in the development of landscape varies according to the phenomena under investigation. Landslides will only occur in response to events of sufficient magnitude to exceed the yield stresses of the rock or regolith. On the other hand stream regimes and sediment yields respond to events of varying magnitude, and according to Wollman and Miller (1960), those of intermediate intensity and frequency will in aggregate produce the most important results.

All these considerations are relevant to the nature of inheritance, and to the methodology of landform analysis. It is within such a context that the use of contemporary averages of both inputs to (such as rainfall) and outputs from (such as discharge and sediment yield) the denudation system should be viewed. Similarly, these considerations must influence the interpretation of morphometric data, and because the form of the landsurface offers few possibilities for the understanding of inheritance, the importance of sub-surface materials is once more emphasised.

The concept of continuity

Paradoxically, it is the continuity of morphogenesis which is one of the most striking outcomes from pedological and geomorphological study of the old lands in the tropics and sub-tropics. This does not derive from a crude morphoclimatic interpretation of tropical landforms, but involves the elaboration of an oscillatory model, in which periodic stability and instability of the groundsurface is seen to be superimposed upon a prolonged sub-aerial development, during which major interruptions from tectonic or climatic causes have been very localised in time and space.

The continuous operation of the soil and weathering systems throughout geological time (Nikiforoff 1959) must be viewed alongside the fluctuating rates of surface denudation. The interaction of these two systems is responsible for the juxtaposition of contemporary regoliths and relict weathering profiles, rock outcrops and areas of deposition. The complexity of such patterns necessarily limits the applications of existing models of landform development which are based upon very fragmentary field evidence. However, knowledge of the inter-tropical zone has now increased to a level that makes clear the limitations of many simple generalisations concerning landform evolution. Cyclic pediplanation, for example, remains a relevant concept, but one capable only of describing the gross form of the continents.

The closer focus of detailed study elucidates the effects of more recent events upon smaller areas, producing a complex mosaic of forms and deposits that can obscure the broader framework of the landscape. In attempts to arrive at a fuller understanding of tropical landforms it is therefore necessary to vary the level of resolution according to the scale and nature of the enquiry.

However, the energy systems which influence landform are global, and both crustal upheaval and climatic change must finally be viewed in this context. Only in this way can local sequences and patterns of deposits and landforms become integrated with the broader concepts of continental landform development. Thus, in this book many individual studies have been used to illustrate general concepts of landform development, although each pertains only to a small field area. Some hazards in this approach will be evident, but only further painstaking field research will enable us to sift the local detail in order to reveal more clearly the fundamental principles of landform development in warm climates.

Bibliography

AHNERT, F. (1970). 'Functional relationships between denudation, relief and uplift in large mid-latitude drainage basins', *Am. J. Sci.*, **268,** 243–263.

AITCHESON, G. D. and GRANT, K. (1970). 'Engineering significance of silcretes and ferricretes in Australia', *Engng. Geol.,* **4,** 93–120.

ALEXANDER, L. T. and CADY, J. G. (1962). 'Genesis and hardening of laterite in soils', *U.S. Dept Agric, Soil Conserv. Serv., Tech. Bull.,* **1218.**

ANDERSON, D. H. and HAWKES, H. E. (1958). 'Relative mobility of the common elements in weathering of some schist and granite areas', *Geochim. cosmochin. Acta,* **14,** 204–210.

ARCHAMBAULT, J. (1960). 'Les eaux souterraines de L'Afrique Occidentale', *Service L'Hydraulique de L'A.O.F.,* **137.**

AUBREVILLE, A. (1962). 'Savanisation tropicale et glaciations quaternaires', *Adansonia N.S.,* **II,** (1), 16–84.

AUGUSTITHIS, S. S. and OTTERMAN, J. (1966). 'On diffusion rings and spheroidal weathering', *Chem. Geol.,* **1,** 201–209.

BAIN, A. O. N. (1923). 'The formation of inselbergs', *Geol. Mag.* **60,** 97–101.

BAKKER, J. P. (1960). 'Some observations in connection with recent Dutch investigations about granite weathering in different climates and climatic changes', *Z. Geomorph., Suppl.,* **1,** 69–92.

BAKKER, J. P. (1965). 'A forgotten factor in the development of glacial stairways', *Z. Geomorph.,* **9,** (1), 18–34.

BAKKER, J. P. (1967). 'Weathering of granites in different climates', In: P. Macar (ed.), *L'Evolution des versants,* Congr. Coll. L'Univ. Liège, **40,** 51–68.

BAKKER, J. P. and LEVELT, Th. W. M. (1964). 'An enquiry into the problems of a polyclimatic development of peneplains and pediments (etchplains) in Europe during the Senonian and Tertiary Period', *Publ. Serv. Carte Geol. Luxembourg,* **14,** 27–75.

BALK, R. (1937). *The Structural Behaviour of the Igneous Rocks,* Ann Arbor, Michigan p. 177.

BARBIER, R. (1967). 'Nouvelles réflexions sur le problème des "pains de sucre" à propos d'observations dans le Tassili N'Ajjer (Algérie), *Trav. Lab. geol., Grenoble,* **43,** 15–21.

BARNES, J. W. (ed.) (1961). 'Mineral Resources of Uganda', *Bull. Geol. Surv. Uganda,* **4.**

BECKETT, P. H. T. and WEBSTER, R. (1965). *A Classification System for Terrain,* Military Engineering Experimental Extablishment, Christchurch, U.K., Rep **872.**

BERRY, L. and RUXTON, B. P. (1959). 'Notes on weathering zones and soils on granite rocks in two tropical regions', *J. Soil Sci.,* **10,** 54–63.

BERTALLANFFY, J. von (1950). 'An outline of general systems theory', *Br. J. Phil. Sci.,* **1,** 134–165.

BIROT, P. (1958). 'Les domes crystallines', Centre Nationale de la Recherche Scientifique (C.N.R.S.), *Mem. Documents,* **6,** 8–34.

BIROT, P. (1962). *Contribution à l'étude de la désagrégation des roches,* Centre Documentation Universitaire, Paris, p. 232.

BIROT, P. (1968). 'The cycle of erosion in different climates', Batsford, London. (Translated from: *Le cycle d'érosions sou les différents climats,* University of Brazil, 1960).

BIROT, P. and DRESCH, J. (1966). 'Pédiments et glacis dans l'ouest des Etats-Unis', *Ann. Géograph.,* **75,** 513–522.

BISDOM, E. B. A. (1967). 'The role of microcrack systems in the spheroidal weathering of an intrusive granite in Galicia (N.W. Spain)', *Geol.- mijnbouwk,* **46,** 333–340.

BISHOP, W. W. (1966). 'Stratigraphic geomorphology: a review of some East African landforms', In: G. H. Dury (ed), *Essays in Geomorphology,* Heinemann, 139–176.

BISHOP, W. W. and CLARK, J. D. (ed), (1967). *Background to Evolution in Africa,* University of Chicago Press, Chicago.

BISSETT, C. B. (1941). 'Water boring in Uganda 1920–1940', *Geol. Surv. Uganda, Water Supply Paper,* **1.**

BLEACKLEY, D. (1964). 'Bauxites and laterites of British Guiana', *Bull. geol. Surv. Br. Guiana,* **34.**

BOND, G. (1949). A preliminary account of the Pleistocene geology of the Plateau tin fields region of Northern Nigeria, *Proc. 3rd West African Conf., Ibadan,* 187–201.

BORNHARDT, W. (1900). *Zur Oberflächengestellung und Geologie, Deutsch Ostafrikas,* Berlin.

BOYE, M. and FRITSCH, P. (1973). 'Dégagement artificiel d'un dôme crystallin au Sud-Cameroun', *Travaux et Documents de Géographie Tropicale* (Bordeaux), **8,** 31–62.

BOYE, M. and SEURIN, M. (1973). 'Les modalités de l'alteration à la carrière d'Ebaka (Sud-Cameroun)', *Travaux et Documents de Géographie Tropicale* (Bordeaux), **8,** 65–94.

BRADLEY, W. C. (1963). 'Large scale exfoliation in massive sandstone of the Colorado Plateau', *Bull. geol. Soc. Am.,* **74,** 519–528.

BRANNER, J. C. (1896). 'Decomposition of rocks in Brazil', *Bull. geol. Soc. Am.,* **7,** 255–314.

BROSCH, A. (1970). 'Observations on the geomorphic relationships of laterite in south-eastern Ankole (Uganda)', *Jerusalem Stud. Geogr.,* **1,** 153–179.

BROWNE, W. R. (1964). 'Grey Billy and the age of tor topography in Monaro, N.S.W.', *Proc. Linnean Soc. N.S.W.*, **89**, (3), 322–325.
BRÜCKNER, W. D. (1955). 'The mantle rock (laterite) of the Gold Coast and its origin', *Geol. Rundschau*, **43**, 307–327.
BRUNSDEN, D. (1964). 'The origin of decomposed granite on Dartmoor', In: I. G. Simmons (ed.) *Dartmoor Essays*, Devon Assoc. Advan. Sci. Lit. Arts, 97–116.
BÜDEL, J. (1957). 'Die *Doppelten Einebnungsflächen* in den feuchten Tropen, *Z. Geomorph.*, N.F., **1**, 201–288.
BÜDEL, J. (1965). 'Die relieftypen der Flächenspülzone: Süd-Indiens am Ostabfall Dekans gegen Madras', *Colloquium Geogr.*, **8**, Bonn.
BÜDEL, J. (1968). 'Geomorphology–Principles', In: R. W. Fairbridge (ed.), *Encyclopaedia of Geomorphology*, Reinhold, New York, 416–422.
BÜDEL, J. (1970). 'Pedimente, Rumpflächen und Rückland–Steilhänge', *Z. Geomorph. N.F.*, **14**, 1–57.
BURKE, K. and DUROTYE, B. (1971). 'Geomorphology and superficial deposits related to late Quaternary climatic variation in south western Nigeria', *Z. Geomorph. N.F.*, **15**, 430–444.
BUTLER, B. E. (1959). *Periodic phenomena in landscapes as a basis for soil studies,* C.S.I.R.O. Aust. Soil Publ., **14**, Canbera.
BUTLER, B. E. (1967). 'Soil periodicity in relation to landform development in south eastern Australia', In: J. N. Jennings and J. A. Mabbutt (eds.), *Landform Studies from Australia and New Guinea*, University Press, Cambridge.
CAMPBELL, J. M. (1917). 'Laterite', *Min. Mag.*, **17**, 67–77 (Pt. 1); 120–128 (Pt. II): 171–179 (Pt. III): 220–229 (Pt. IV).
CARROLL, Dorothy (1962). *Rainwater as a chemical agent of denudation —A review,* U.S. Geol. Surv. Water Supply Paper, 1535—G.
CARROLL, Dorothy (1970). *Rock Weathering*, Plenum Publications, New York.
CARSON, M. A. and KIRKBY, M. J. (1972). '*Hillslope Form and Process*', University Press, Cambridge.
CHALMERS, R. (1898). 'The pre-glacial decay of rocks in eastern Canada', *Am. J. Sci.,* 4th Ser., **V**, 273–282.
CHAPMAN, C. A. (1958). 'The control of jointing by topography', *J. Geol.* **66**, 552–558.
CHAPMAN, C. A. and RIOUX, R. L. (1958). 'Statistical study of topography, sheeting and jointing in granite, Acadia National Park, Maine', *Am. J. Sci.,* **256**, 111–127.
CHORLEY, R. J. (1957). 'Climate and morphometry', *J. Geol.*, **65**, 628–638.
CHORLEY, R. J. (1962). 'Geomorphology and general systems theory' *Bull. U.S. geol. Surf.,* Professional Paper, 500–B.
CHORLEY, R. J. and MORGAN, M. A. (1962). 'Comparison of morphometric features, Unaka Mountains, Tennessee and North Carolina, and Dartmoor, England', *Bull. geol, Soc. Am.,* **73**, 17–34.
CHRISTIAN, C. S. (1957). 'The concept of land units and land systems', *Proc. 9th Pacific Sci. Conf.*, **20**, 74–81.

CHURCHWARD, H. M. (1969). 'Erosional modification of a lateritised landscape over sedimentary rocks. Its effects on soil distribution', *Aust. J. Soil Res,* **8,** 1–19.

CLAYTON, R. W. (1956). 'Linear depressions (Bergfüssneiderungen) in savanna landscapes', *Geogrl. Studies,* **3,** 102–126.

CLAYTON, W. D. (1958). 'Erosion surfaces in Kabba Province, Nigeria', *J. West African Sci. Assoc.,* **4,** 141–149.

CLAYTON, K. M. (1970). 'The problem of field evidence in geomorphology', In: Osborne, R. H., Barnes, F. A. and Doornkomp, J. C. (eds.), *Geographical Essays in Honour of Professor K. C. Edwards,* University of Nottingham, Nottingham, 131–139.

COETZEE, J. A. (1964). 'Evidence for a considerable depression of the vegetation belts during the Upper Pleistocene on the East African mountains', *Nature,* **204,** 564–566.

CONNAH, T. H. and HUBBLE, G. D. (1962). 'Laterites', In: D. Hill and A. K. Denmead (eds.), *The Geology of Queensland,* Melbourne, 373–386.

COOKE, R. U. (1970). 'Morphometric analysis of pediments and associated landforms in the western Mojave Desert, California', *Am. J. Sci.,* **269,** 26–38.

CORBEL, J. (1957). 'L'érosion climatique des granites et silicates sous climats chauds', *Rev. Géomorph. Dyn.,* **8,** 4–8.

CORBEL, J. (1959). 'Vitesse de L'érosion', *Z. Geomorphol.,* **3,** 1–28.

CORBEL, J. (1964). 'L'érosion terrestre, étude quantitative (methodes – techniques – résultats)', *Ann. Géograph,* **73,** 385–412.

CORRENS, C. W. (1963). 'Experiments on the decomposition of silicates and discussion of chemical weathering', *Clays, Clay Miner.* V, **12,** 443–460.

COTTON, C. A. (1917). 'Block mountains in New Zealand', *Am. J. Sci.,* **194,** 249–293. *N.Z. Jl. Geol. Technol.* B, **20,** 1–8.

COTTON, C. A. (1942). *Climatic Accidents in Landscape Making,* Whitcomb and Tombs, Wellington.

COTTON, C. A. (1961). 'Theory of Savanna planation', *Geography,* **46,** 89–96.

COTTON, C. A. (1962a). 'The origin of feral (fine textural) relief', *N.Z. Jl. Geol. Geophys.,* **5,** 269–270.

COTTON, C. A. (1962b). 'Plains and inselbergs of the humid tropics'. *Roy. Soc. New Zealand Trans.* (*Geology*), **1,** (18), 269–277.

CRAIG, D. C. and LOUGHNAN, F. C. (1964). 'Chemical and numerological transformations accompanying the weathering of basic volcanic rocks from New South Wales', *Aust. J. Soil Res.,* **2,** 218–234.

CREDNER, W. (1931). 'Das Kräfteverhältnis morphogenetischer Faktoren und ihr Ausdruck im Formenbild Südostasiens', *Bull. geol. Soc. China,* Peiking, **11,** 13–34.

CUNNINGHAM, F. F. (1969). 'The Crow Tors, Laramie Mountains, Wyoming, U.S.A.', *Z. Geomorph.,* N.F., **13,** 56–74.

CUNNINGHAM, F. F. (1971). 'The Silent City of Rocks, a bornhardt landscape in the Cotterel Range, South Idaho, U.S.A.', *Z. Geomorph.*, N.F., **15**, (4), 404–429.

DALE, T. N. (1923). 'The commercial granites of New England', *Bull. U.S. geol. Surv.*, **738**.

DAVIS, S. N. (1964). 'Silica in streams and ground water', *Am. J. Sci.*, **262**, 870–891.

D'HOORE, J. L. (1954). *L'accumulation des sesquioxides libres dans les sols tropicaux, Publ. I.N.E.A.C. Sér. Sci.*, **62**, Brussels.

D'HOORE, J. L. (1964). *Soil Map of Africa scale 1:5 000 000 Explanatory Monograph, C.C.T.A., Publ.*, **93**, Lagos.

De LAPPARENT, U. (1941). 'Logique des minéraux du granite', *Rev. Scientif.*, 284–292.

De MEIS, M. R. M. and da SILVA, J. X. (1968). 'Mouvement de masses-récentes à Rio de Janeiro: Une étude de géomorphologie dynamique', *Rev. Géomorph. Dyn.*, **18**, (4), 1–15.

DEMEK, J. (1964). 'Slope development in granite areas of the Bohemian massif, Czechoslovakia', *Z. Geomorph.*, Suppl., **5**, 82–106.

DENNEN, W. H. and ANDERSON, P. J. (1962). 'Chemical changes in incipient rock weathering', *Bull. geol. soc. Am.*, **73**, 375–384.

DEPETRIS, P. J. and GRIFFIN, J. J. (1968). Suspended load in the Rio de la Plata drainage basin, *Sedimentology*, **11**, 53–60.

DE SWARDT, A. M. J. (1946). The recent erosional history of the Kaduna Valley, near Kaduna township, *Rep. geol. Surv., Nigeria.* 39–45.

DE SWARDT, A. M. J. (1949). 'Recent erosion surfaces on the Jos Plateau', *Proc. Third West African Conf., Ibadan*, 180–186.

DE SWARDT, A. M. J. (1953). 'The Geology of the country around Ilesha', *Bull. geol. Surv. Nigeria*, **23**.

DE SWARDT, A. M. J. (1964). 'Lateritisation and landscape development in equatorial Africa', *Z. Geomorph.*, N.F., **8**, 313–333.

DE SWARDT, A. M. J. and CASEY, O. P. (1963). 'The Coal Resources of Nigeria', *Bull. geol. Surv. Nigeria*, **28**.

DE SWARDT, A. M. J. and TRENDALL, A. F. (1969). 'The physiographic development of Uganda', *Overseas Geol. Miner. Resour.*, (Gt. Br.), **10**, (3), 241–288.

DE VILLIERS, J. M. A. (1965). 'Present soil forming factors and processes in tropical and subtropical regions', *Soil Sci.*, **99**, 50–57.

DIXEY, F. (1931). *A Practical Handbook of Water Supply*, Murty, London.

DIXEY, F. (1955). 'Erosion surfaces in Africa: some considerations of age and origin', *Trans. Proc. Geol. Soc. South Africa*, **58**, 265–280.

DOORNKAMP, J. C. (1968). 'The role of inselbergs in the geomorphology of southern Uganda', *Trans. Inst. Br. Geogr.*, **44**, 151–162.

DOORNKAMP, J. C. (1970). 'The geomorphology of the Mbarara Area' - Sheet SA-36-1, *Geomorphological Rep.* **No. 1**, Dept. Geogr. Univ. Nottingham and Geol. Surv. Mines Dep., Uganda.

DOORNKAMP, J. C. and KING, C. A. M. (1971). *Numerical Analysis in Geomorphology*, Arnold, London.

DOUGLAS, I. (1967a). 'Natural and man-made erosion in the humid tropics of Australia, Malaysia and Singapore', In: *Symposium on River Morphology*, General Assembly of Bern.

DOUGLAS, I. (1967b). 'Man, vegetation and the sediment yields of rivers', *Nature,* **215,** (No. 5104), 925–928.

DOUGLAS, I. (1968). 'Erosion of granite terrains under tropical rain forest in Australia, Malaysia and Singapore', In: *Symposium on, River Morphology*, General Assembly, Bern, 1967.

DOUGLAS, I. (1969). 'The efficiency of humid tropical denudation systems', *Trans. Inst. Br. Geogr.*, **40,** 1–16.

DOWNING, A. (1968). 'Subsurface erosion as a geomorphic agent in Natal', *Trans. geol. Soc. South Africa,* **71,** 131–134.

DRESCH, J. (1953). 'Plaines Soudanaises', *Rev. Géomorph. dyn.*, **4,** 39–44.

DRISCOLL, E. M. (1964). 'Landforms in the Northern Territory of Australia', In: R. W. Steel and R. M. Prothero (eds), *Geographers and the tropics*, Longmans, London.

DU PREEZ, J. W. (1949). 'Laterite: a general discussion with a description of Nigerian occurrences', *Bull. Agric. Congo Belge,* **40,** (1), 53–66. (Also published in similar form: 'Origin, Classification and Distribution of Nigerian Laterites', *Proc. 3rd West African Conf.*, Ibadan, 1949, 223–234).

DURY, G. H. (1967), (ed.). *Essays in Geomorphology*, Heinemann, London, p. 404.

DURY, G. H. (1969). 'Rational descriptive classification of duricrusts', *Earth Sci. J.,* **3,** 77–86.

DURY, G. H. (1971). 'Relict deep weathering and duricrusting in relation to the palaeoenvironments of middle latitudes', *Geogrl. J.,* **137,** 511–522.

EDEN, M. J. (1971). 'Some aspects of weathering and landforms in Guyana (formerly British Guiana)', *Z. Geomorph.*, N.F., **15,** (2), 181–198.

EDEN, M. J. and GREEN, C. P. (1971). 'Some aspects of granite weathering and tor formation on Dartmoor, England', *Geografiska Annaler,* **53A,** 92–99.

EGGLER, D. H., LARSON, E. E. and BRADLEY, W. C. (1969). 'Granites, grusses and the Sherman erosion surface, southern Laramie Range, Colorado – Wyoming', *Am. J. Sci.,* **267,** 510–522.

ENSLIN, J. F. (1961). 'Secondary aquifers in South Africa and the scientific selection of boring sites in them', C.C.T.A., *Intern. African Conf. Hydrol.,* Nairobi, Publ. **66** Sect. 4, Ground Water Hydrology, 379.

ERHART, H. (1955). '*Biostasie* et *Rhexistasie:* Esquisse d'une théorie sur le rôle de la pédogenèse en tant que phénomène géologique', *Compt. Rend Acad. Sci. Francaise,* **241,** 1218–1220.

ERHART, H. (1956). *La genèse des sols en tant que phénomène géologique*, Masson, Paris.

EVANS, I. S. (1971). 'Salt crystallisation and rock weathering: a review', *Rev. geomorph. Dyn.,* **19,** (4), 153–171.

EYLES, R. J. (1968). 'Stream net ratios in West Malaysia', *Bull. geol. Soc. Am.*, **79**, 701–712.
EYLES, R. J. (1969). 'Depth of dissection of the West Malaysian landscape, *J. Trop. Geogr.*, **28**, 23–31.
EYLES, R. J. (1971). 'A classification of West Malaysian drainage basins', *Annals Ass. Am. Geogr.*, **61**, 460–467.
EYLES, R. J. and HO, R. (1970). 'Soil creep on a humid tropical slope', *J. Trop. Geogr.*, **31**, 40–42.
FALCONER, J. D. (1911). *The Geology and Geography of Northern Nigeria*, Macmillan, London.
FALCONER, J. D. (1912). 'The origin of kopjes and inselbergs', *Br. Assoc. Adv. Sci. Trans.*, Sect. C., **476.**
FAURE, H. (1967). 'Lacs quaternaires du Sahara', *Intern. Symp. Paleolininology*, Tihany.
FEININGER, T. (1971). 'Chemical weathering and glacial erosion of crystalline rocks and the origin of till', In: *Geological Survey Research* 1971, U.S. Geol. Surv., Prof. Pap. **750-C,** C65–C81.
FIELDES, M. and SWINDALE, L. D. (1954). 'Chemical weathering of silicates in soil formation', *J. Sci. Techn.*, New Zealand, **56**, 140–154.
FITZPATRICK, E. A. (1963). 'Deeply weathered rock in Scotland, its occurrence, age and contribution to the soils', *J. Soil Sci.*, **14,** 33–43.
FLACH, K. W., CADY, J. G. and NETTLETON, W. D. (1968). 'Pedogenic alteration of highly weathered parent materials', *Ninth Intern. Congr. Soil Sci. Trans.*, **4,** 343–351.
FLOHN, H. (1963). 'Zur meteorologischen interpretation der pleistozanen Klimaschwankungen', *Eiszeitalter U. Gegenwart*, **14**, 153–160.
FÖLSTER, H. (1964). 'Morphogenese der südsudanesichen Pediplane', *Z. Geomorph.*, N.F., **8**, 393–423
FÖLSTER, H. (1969). 'Slope development in SW-Nigeria during late Pleistocene and Holocene', *Gottinger Bodenkundliche Berichte*, **10,** 3–56.
FOURNIER, F. (1960). *Climat et érosion: la relation entre l'érosion du sol par l'eau et les précipitiatons atmosphériques*, Presses Univ., Paris.
FOURNIER, F. (1962). *Carte du danger d'érosion en Afrique au Sud du Sahara*, C.E.E.–C.C.T.A., Presses Univ., Paris.
FRANKEL, J. J. and KENT, L. E. (1937). Grahamstown surface quartzites (silcretes). *Trans. Geol. Soc. S. Africa*, **40,** 1–52.
FREISE, F. W. (1938). 'Inselberge und Inselberg-landschaften im granit und gneissgebiet, Brasiliens', *Z. Geomorph.*, **X,** 137–168.
FRIPIAT, J. J. and HERBILLON, A. J. (1971). 'Formation and transformation of clay minerals in tropical soils', In: *Soils and Tropical Weathering, Proc. Bandung Symp.*, UNESCO 1969, 15–24.
GELLERT, J. F. (1970). 'Climatomorphology and palaeoclimates of the central European Tertiary', In: M. Pecsi (ed.), *Problems of Relief Planation*, Akadémiai kaidó, Budapest, 107–112.
GERBER, E. and SCHEIDEGGER, A. E. (1969). 'Stress induced weathering of rock messes', *Ecologae geol. Helv.*, **62,** (2), 401–414.

GIBBS, R. J. (1967). 'The geochemistry of the Amazon River system, Part I', *Bull. geol. Soc. Am.*, **78**, 1203–1232.

GILBERT, G. K. (1904). 'Domes and domed structures of the high Sierra', *Bull. geol. Soc. Am.*, **15**, 29–36.

GJESSING, J. (1967). 'Norway's Paleic surface', *Norsk geog. Tidsskr.*, **21**, 69–132.

GODARD, A. (1966). 'Morphologie des socles et des massifs anciens' (4th part), *Rev. Geomorph. de L'Est*, **4**, 77–96.

GOLDICH, S. S. (1938). 'A study in rock weathering', *J. Geol.*, **46**, 17–58.

GOUDIE, A. (1970). 'Input and output considerations in estimating rates of chemical denudation', *Earth Sci. J.*, **4**, 59–65.

GOUDIE, A. (1972). 'Duricrusts in Tropical and Subtropical Landscapes', Clarendon, Oxford.

GRANT, W. H. (1969). 'Abrasion pH, an index of chemical weathering', *Clays Clay Miner.*, **17**, 151–155.

GROVE, A. T. (1958). 'The ancient erg of Hausaland and similar formations on the south side of the Sahara', *Geogrl. J.*, **124**, 528–33.

GROVE, A. T. and PULLAN, R. A. (1963). 'Some aspects of the Pleistocene paleogeography of the Chad Basin', In: F. C. Howell and F. Bouliere (eds.), *African Ecology and Human Evolution*, Methuen, London.

GROVE, A. T. and WARREN, A. (1968). 'Quaternary landforms and climate on the south side of the Sahara', *Geogrl. J.*, **134**, (2), 193–208.

GROVE, A. T. (1972). 'The dissolved and solid load carried by some west African rivers: Senegal, Niger, Benue and Shari', *J. Hydrol.*, **XVI**, 277–300.

GRUBB, P. L. C. (1968). *Geology and Bauxite Deposits of the Pengarang Area, Southeast Johore,* Geol. Surv. West Malaysia, District Mem., **14**,

HAANTJENS, H. A. (1965). 'Practical Aspects of Land System Surveys in New Guinea', *J. Trop. Geogr.*, **21**, 12–20.

HACK, J. T. (1960). 'Interpretation of erosional topography in humid temperate regions', *Am. J. Sci.*, **258-A**, 80–97.

HACK, J. T. (1966). 'Circular patterns and exfoliation in crystalline terraine, Grandfather Mountain area, North Carolina', *Bull. geol. Soc. Am.*, **77**, 975–987.

HAMBLER, D. J. (1964). 'The vegetation of granite outcrops in western Nigeria', *J. Ecol.*, **52**, 573–594.

HANDLEY, J. R. F. (1952). 'The geomorphology of the Nzega area of Tanganyika with special reference to the formation of granite tors', *Congr. geol. Intern. Compt. Rend. 21e*, Algiers, 201–210.

HARLAND, W. B. (1957). 'Exfoliation joints and ice action', *J. Glaciol.*, **3**, 8–10.

HARPUM, J. R. (1963). 'Evolution of granite scenery in Tanganyika', *Geol. Surv. Tanganyika Rec.*, **10**, 39–46.

HARRIS, R. C. and ADAMS, J. A. S. (1966). 'Geochemical and Mineralogical Studies on the Weathering of Granite Rocks', *Am. J. Sci.*, **264**, (2), 146–173.

HARRISON, J. B. (1934). *The katamorphism of igneous rocks under humid tropical conditions,* Imperial Bureau Soil Sci., Harpenden, England.

HAYS, J. (1967). 'Land surfaces and laterites in the north of the Northern Territory', In: J. N. Jennings and J. A. Mabbutt (eds.), *Landform Studies from Australia and New Guinea*, University Press, Cambridge.

HOLEMAN, J. N. (1968). 'The sediment yield of major rivers of the world', *Water Resources Res.,* **4,** 737–747.

HOLMES, A. and WRAY, A. (1913). 'Mozambique: a geographical study', *Geogrl. J.,* **62**, 143–152.

HUNTER, J. M. and HAYWARD, D. F. (1971). 'Towards a mode of scarp retreat and drainage evolution: Ghana, West Africa', *Geogrl. J.,* **137,** (1), 51–68.

HURAULT, J. (1967). *L'érosion régressive dans les régions tropicales humides et la genèse des inselbergs granitiques,* Etudes Photo-interpretation **3**, Inst. Geograph. Natl. Paris.

JAHNS, R. H. (1943). 'Sheet structure in granites: its origin and use as a measure of glacial erosion', *J. Geol.,* **51,** 71–98.

JEJE, L. K. (1970). *Some Aspects of the Geomorphology of south western Nigeria,* Ph.D. thesis. (University of Edinburgh).

JEJE, L. K. (1972). 'Landform development at the boundary of sedimentary and crystalline rocks in south western Nigeria, *J. Trop. Geogr.,* **34,** 25–33.

JENNINGS, J. N. (1967). 'Some karst areas of Australia', In: J. N. Jennings and J. A. Marbutt (eds.), *Landform Studies from Australia and New Guinea,* University Press, Cambridge.

JENNINGS, J. N. and BIK, M. J. (1962). 'Karst morphology in Australian New Guinea', *Nature,* **194,** 4833, 1036–1038.

JENNY, H. (1941). *Factors in Soil Formation,* McGraw-Hill, New York.

JESSEN, O. (1936). '*Reisen und Forschingen in Angola,* Berlin.

JESSEN, O. (1938). 'Tertiarklima und Mittelgebirgsmorphologie', *Z. Geo. fur Erdk.,* Berlin.

JESSUP, R. W. (1960). 'The lateritic soils of the south eastern portion of the Australian arid zone', *J. Soil Sci.,* **11,** 92–105.

JONES, T. R. (1859). 'Notes on some granite tors', *Geologist,* **2,** 301–312.

KAITANEN, V. (1969). 'A geographical study of the morphogenesis of northern Lapland', *Fennia,* **99,** (5).

KAYE, C. A. (1967). 'Kaolinisation of bedrock of the Boston: Massachusetts, area', *Bull. U.S. geol. Surv., Professional Paper,* **575-C,** C165–C172.

KAYSER, K. (1957). 'Zur Flächenbildung, Stufen – und Inselberg – Entwicklung in den wechselfeuchten Tropen auf der Ostseite Süd-Rhodesiens', *Tagungsbericht und wiss. Abhandlungen, Deutscher Geogr. Aphentag.,* Wurzburg, 1957. 165–172.

KAYSER, K. und OBST, E. (1949). *Die gross Randstufe auf der Ostseite Südafrikas und ihr Vorland,* Sonderveröffentlichungen der Geogr. Gesellschaft, Hanover,

KEAY, R. W. J. (1959). 'Derived savannas, derived from what?', *Bull. Inst. Française Afrique Noire*, Ser. A, **2,** 427–438.

KELLER, W. D. (1957). *The Principles of Chemical Weathering*, Lucas Brothers, Missouri.

KIESLINGER, A. (1960). 'Residual stress and relaxation in rocks', *Intern. Geol. Congr. Copenhagen, 1960, Rept. Session,* **17,** 270–276, Norden.

KING, L. C. (1948). 'A Theory of bornhardts', *Geogrl. J.,* **112,** 83–87.

KING, L. C. (1953). 'Canons of landscape evolution', *Bull. geol. Soc. Am.,* **64,** 721–752.

KING, L. C. (1957). 'The uniformitarian nature of hillslopes', *Trans. Edinburgh geol. Soc.,* **17,** 81–102.

KING, L. C. (1958). 'Correspondence: the problems of tors', *Geogrl. J.,* **124,** 289–291.

KING, L. C. (1962). *The Morphology of the Earth*, Oliver and Boyd, Edinburgh.

KING, L. C. (1966). 'The origins of bornhardts', *Z. Geomorph.* N.F., **10,** 97–98.

KIRKBY, M. J. (1967). 'Movement and theory of soil creep', *J. Geol.,* **75,** 359–378.

KIRKBY, M. J. (1969). 'Erosion by water on hillslopes', In: R. J. Chorley (ed.), *Water, Earth and Man*, Methuen, London.

KNILL, J. L. and JONES, K. S. (1965). 'The recording and interpretation of geological conditions in the foundations of the Roseires, Karriba and Latiyan Dams', *Geotechnique*, **15,** 94–124.

KRAUSKOPF, K. B. (1959). *Introduction to Geochemistry*, McGraw Hill, New York.

LAMOTTE, M. and ROUGERIE, G. (1962). 'Les apports allochtones dans la genèse de cuirasses ferrugineuses', *Rev. Geomorph. Dyn.,* **13,** 145–160.

LANGBEIN, W. B. and SCHUMM, S. A. (1958). 'Yield of sediment in relation to mean annual precipitation', *Trans. Am. geophys. Union,* **39,** 1076–1084.

LANGFORD-SMITH, T. and DURY, G. H. (1965). 'Distribution, character and attitude of the duricrust in the north west of New South Wales and the adjacent areas of Queensland', *Aus. J. Sci.,* **263,** 170–190.

LEDGER, D. L. (1964). 'Some hydrological characteristics of West African rivers', *Trans. Inst. B. Geogr.,* **35,** 73–90.

LEDGER, D. L. (1969). 'Dry season flow characteristics of West African rivers', In: M. F. Thomas and G. Whittington (eds), *Environment and Land Use in Africa*, Methuen, London.

LELONG, F. (1966). 'Régime des nappes phréatiques contenues dans les formations d'altération tropicale. Conséquences pour la pédogenèse', *Sci. Terre,* **11,** 203–244.

LELONG, F. and MILLOT, G. (1966). 'Sur L'origine des mineraux micaces des altérations latéritiques. Diagenèse Régressive – Minéraux en transit', *Bull. Ser. Carte. geol. Alsace Lorraine,* **19,** 271–287.

LENEUF, N. and AUBERT, G. (1960). 'Essai d'évaluation de la vitesse de ferrallitisation', *Proc. 7th Int. Congr. Soil Sci.*, 225–228.

LEOPOLD, L. B., WOLMAN, M. G. and MILLER, J. P. (1964). *Fluvial Processes in Geomorphology*, Freeman, San Francisco.

LINTON, D. L. (1955). 'The problem of tors', *Geogrl. J.*, **121**, 470–487.

LINTON, D. L. (1958). 'Correspondence: the problem of tors', *Geogrl. J.*, **124**, 289–291.

LIVINGSTONE, D. A. (1963). 'Chemical composition of rivers and lakes', *O.S. geol. Surv.*, Professional Paper, 440–G.

LOUGHNAN, F. C. (1969). *The Chemical Weathering of the Silicate Minerals*, Elsevier, New York, p. 154.

LOUIS, H. (1964). 'Über rumpfflächen – und talbildung in den wechselfeuchten tropen besonders nach studien in Tanganyika', *Z. Geomorph.*, **8**, (Sonderheft), 43–70.

LOVERING, T. S. (1959). 'Geologic significance of accumulator plants in rock weathering', *Bull. geol. Soc. Am.*, **70**, 781–800.

LUMB, P. (1962). 'The properties of decomposed granite', *Geotechnique*, **12**, 226–243.

LUMB, P. (1965). 'The residual soils of Hong Kong', *Geotechnique*, **15**, (2), 180–194.

MABBUTT, J. A. (1952). 'A study of granite relief from South West Africa', *Geol. Mag.*, **89**, 87–96.

MABBUTT, J. A. (1961a). 'A stripped landsurface in Western Australia', *Trans. Inst. Br. Geog.* **29**, 101–114.

MABBUTT, J. A. (1961b). ' "Basal Surface" or "Weathering Front" ', *Proc. Geol. Assoc.*, **72**, 357–359.

MABBUTT, J. A. (1962). 'Geomorphology of the Alice Springs area', In: R. A. Perry *et al.* 1962, *Lands of the Alice Springs Area, Northern Territory*, 1956–57, C.S.I.R.O. Australia Land Res. Series, **8**, 163–178.

MABBUTT, J. A. (1965). 'The weathered landsurface of central Australia', *Z. Geomorph.*, N.F., **9**, 82–114.

MABBUTT, J. A. (1966). 'The mantle controlled planation of pediments', *Am. J. Sci.*, **264**, 78–91.

MABBUTT, J. A. and SCOTT, R. M. (1966). 'Periodicity of morphogenesis and soil formation in a savannah landscape near Port Moresby, Papua', *Z. Geomorph.*, N.F., **10**, 69–89.

McCALL, G. J. H., BAKER, B. H. and WALSH, J. (1967). 'Late Tertiary and Quaternary sediments in the Kenya Rift Valley', In: W. W. Bishop and J. D. Clark (eds), *Background to Evolution in Africa*, University of Chicago, Chicago.

McCALLIEN, W. J., RUXTON, B. P. and WALTON, B. J. (1964). 'Mantle rock tectonics: a study in tropical weathering at Accra, Ghana', *Overseas geol. Mineral Resources*, **9**, (3), 257–294.

McCONNELL, R. B. (1968). 'Planation surfaces in Guyana', *Geogrl. J.*, **134**, (4), 506–520.

McGEE, W J. (1897). 'Sheetflood erosion', *Bull. geol. Soc. Am. Bull.*, **8**, 87–112.

MACKIN, J. H. (1970). 'Origin of pediments in the western United States', In: M. Pecsi (ed.), *Problems of Relief Planation,* Budapest.

MAIGNIEN, R. (1958). *Contribution à l'étude du cuirassement des sols en Guinée française,* Serv. Carte. geol. Alsace Lorraine. Mem., **16.**

MAIGNIEN, R. (1960a). 'Influences anciennes sur la morphologie, l'évolution et la répartition des sols en Afrique tropicale de l'Ouest', *7th Intern. Congr. Soil Sci.,* Madison, 171–176.

MAIGNIEN, R. (1960b). 'Soil cuirasses in tropical Africa', *African Soils,* **4,** (4), 5–42.

MAIGNIEN, R. (1966). *Review of Research on Laterites,* UNESCO Natural Resources Res. IV, Paris.

MATTHES, F. E. (1930). 'The geologic history of the Yosemite Valley,' *U.S. Geol. Surv.,* Professional Paper, **160.**

MAUD, R. R. (1965). 'Laterite and lateritic soil in coastal Natal, South Africa', *J. Soil Sci.,* **16,** 60–72.

MAUD, R. R. (1968). 'Further observations on the laterites of coastal Natal, South Africa', *9th Int. Congr. Soil Sci.,* Adelaide, Trans. Vol. **IV,** 151–158.

MELTON, M. A. (1957). *An analysis of the relations among elements of climate, surface properties, and geomorphology,* Office Naval Res. Project NR 389–042 Tech. Rept. **11,** New York.

MELTON, M. A. (1965). 'Debris covered hillslopes of the southern Arizona desert – consideration of their stability and sediment contribution', *J. Geol.,* **73,** 715–729.

MENDELSSOHN, F. (1961). *The Geology of the Northern Rhodesian Copperbelt*, Macdonald, London.

MENSCHING, H. (1970). Planation in arid, subtropic and tropic regions, In: M. Pecsi (ed.), *Problems of Relief Planation* (Studies in Geography in Hungary 8), 73–84.

MICHEL, P. (1967). *Les grandes étapes de la morphogenèse dans les bassins des fleuves Sénégal et Gambie pendant le quaternaire,* 6th Pan African Congr. Prehist. Quatern., Dakar.

MILLOT, G. (1970). *Geology of Clays*, Masson, Paris.

MOHR, E. C. and VAN BAREN, F. A. (1954). *Tropical Soils*, Interscience Publishers, London.

MOSS, R. P. (1965). 'Slope development and soil morphology in a part of south west Nigeria', *J. Soil Sci.,* **16,** 192–209.

MOSS, R. P. (1968): 'Soils, slopes and surfaces in tropical Africa', In: R. P. Moss (ed.), *Soil Resources of Tropical Africa*, University Press, Cambridge.

MOSS, R. P. (1969). 'The ecological background to land-use studies in tropical Africa, with special reference to the West', In: M. F. Thomas and G. Whittington (eds), *Environment and Land Use in Africa*, Methuen, London.

MOUNTAIN, E. D. (1951). 'The origin of silcrete', *South African J. Sci.,* **48,** 201–204.

MULCAHY, M. J. (1959). 'Topographic relationships of laterite near York, W.A.', *J. Roy. Soc. W. Aust.* **42,** 44–48.

MULCAHY, M. J. (1960). 'Laterites and lateritic soils in south western Australia', *J. Soil Sci.,* **11,** 206–225.

MULCAHY, M. J. (1961). 'Soil distribution in relation to landscape development', *Z. Geomorph.,* N.F., **5,** (3), 211–225.

MULCAHY, M. J. (1964). 'Laterite residuals and sandplains', *Aust. J. Sci.,* **27,** 54, 55.

MULCAHY, M. J. (1967). 'Landscapes, laterites and soils in south western Australia', In: J. N. Jennings and J. A. Mabbutt (eds), *Landform Studies from Australia and New Guinea,* University Press, Cambridge.

MULCAHY, M. J., CHURCHWARD, H. M. and DIMMOCK, G. M. (1972). 'Landforms and soils on an uplifted peneplain in the Darling Range, Western Australia', *Aust. J. Soil Res.,* **10,** 1–14.

MULCAHY, M. J. and HINGSTON, F. J. (1961). 'The development and distribution of the soils of the York-Quairaiding Area, Western Australia, in relation to landscape evolution', *C.S.I.R.O., Soil Publ.,* **17,** Canberra.

NAGELL, R. H. (1962). 'The geology of the Sierra do Navio manganese district, Brazil', *Econ. Geol.,* **57,** 481–498.

NIKIFOROFF, C. C. (1949). 'Weathering and soil evolution', *Soil Sci.,* **67,** 219–230.

NIKIFOROFF, C. C. (1959). 'Reappraisal of the soil', *Science,* **129,** 186–196.

NOSSIN, J. J. (1964). 'Geomorphology of the surrounding of Kuanton (Eastern Malaya)', *Geol. Minjbouw.,* **45,** 157–182.

NOSSIN, J. J. (1967). 'Igneous rock weathering on Singapore Island', *Z. Geomorph.,* N.F., **11,** 14–38.

NYE, P. H. (1954). 'Some soil forming processes in the humid tropics. Pt. I: A field study of a catena in the West African forest', *J. Soil Sci.,* **5.** 7–27.

NYE, P. H. (1955). 'Some soil forming processes in the humid tropics. Pt. II: The development of the upper slope member of the catena', *J. Soil Sci.,* **6,** 51–62.

NYE, P. H. and GREENLAND, D. J. (1960). 'The Soil under shifting cultivation', Technical Communication **51,** Commonwealth Bureau of Soils, Commonwealth Agricultural Bureaux, Farnham Royal.

OEN, I. S. (1965). 'Sheeting and exfoliation in the granites of Sermersoq, South Greenland', *Meddelelser Grønland,* **179,** 1–40.

OLLIER, C. D. (1959). 'A two cycle theory of tropical pedology', *J. Soil Sci.,* **10,** 137–148.

OLLIER, C. D. (1960). 'The inselbergs of Uganda', *Z. Geomorph. N.F.,* **4,** 43–52.

OLLIER, C. D. (1965). 'Some features of granite weathering in Australia', *Z. Geomorphol. N.F.,* **9,** 285–304.

OLLIER, C. D. (1967). 'Spheroidal weathering, exfoliation and constant volume alteration', *Z. Geomorph.,* N.F., **9,** 285–304.

OLLIER, C. D. (1969). *Weathering*, Oliver and Boyd, Edinburgh.

OLLIER, C. D. and TUDDENHAM, W. G. (1961). 'Inselbergs of central Australia', *Z. Geomorph.*, **5**, 257–276.

OLLIER, C. D. (1971). 'Causes of spheroidal weathering', *Earth Sci. Rev.*, **7**, 127–141.

ONG, H. Ling, SWANSON, V. E. and BISQUE, R. E. (1970). 'Natural organic acids as agents of chemical weathering', *Geological Survey Research 1970*, U.S. Geol. Surv. Profess. Paper 700–C, C130–C137.

PALLISTER, J. W. (1956). 'Slope development in Buganda', *Geogrl. J.*, **122**, 80–87.

PALMER, J. and NEILSON, R. A. (1962). 'The origin of granite tors on Dartmoor, Devonshire', *Proc. Yorkshire geol. Soc.*, **33**, 315–340.

PASSARGE, S. (1928). *Panoramen afrikanischer; Inselberg – landschaften*, Reimer, Berlin.

PEDRO, G. (1968). 'Distribution des principaux types d'altération chimique a la surface du globe', *Rev. Géogr. Phys. Géol. Dyn.*, **10**, 457–470.

PEEL, R. F. (1966). 'The landscape in aridity', *Trans. Inst. Br. Geog.*, **38**, 1–23.

PELTIER, L. C. (1950). 'The geographical cycle in periglacial regions as it is related to climatic geomorphology', *Am. Assoc. geogr. Ann.*, **40**, 214–236.

PELTIER, L. C. (1962). 'Area sampling for terrain analysis', *Professional Geographer*, **14**, 24–28.

PENCK, W. (1924). *Die Morphologische Analyse*, Stuttgart, English trans. by H. Czech and K. C. Boswell, *Morphological Analysis of Landforms*, Macmillan, London 1953.

PLAYFORD, P. E. (1954). 'Observations on laterite in Western Australia', *Aust. J. Sci.*, **17**, 11–14.

POKORNY, J. (1963), 'The development of mogotes in the southern part of the Cracow Upland', *Bull. Acad. Polonaise Sci., Série des Sci. géol. géogr.*, **XI**, 3, 169–175.

POLYNOV, B. B. (1937). *The Cycle of Weathering* (trans. A. Muir), Murty, London.

PRESCOTT, J. A. and PENDLETON, R. L. (1952). 'Laterites and Lateritic Soils', *Comm. Bur. Soil Sci., Tech. Comm.*, **47**, 51 p.

PUGH, J. C. (1956). 'Fringing pediments and marginal depressions in the inselberg landscape of Nigeria', *Trans. Inst. Br. Geogr.*, **22**, 15–31.

PUGH, J. C. (1966). 'Landforms in low latitudes', In: G. H. Dury (ed.), *Essays in Geomorphology*, Heinemann, London.

PULLAN, R. A. (1967). 'A morphological classification of lateritic ironstones and ferruginised rocks in Northern Nigeria', *Nigerian J. Sci.*, **1**, 161–173.

RAHN, P. H. (1966). 'Inselbergs and nick points in south western Arizona', *Z. Geomorph.*, N.F., **10**, 217–225.

RAPP, A. (1960). 'Development of mountain slopes in Karkervagge and surroundings, northern Scandinavia', *Geogr. Annaler*, **42**, 65–187.

REICHE, P. (1943). 'Graphic representation of chemical weathering', *J. Sediment. Petrol.* **13**, 58–68.

REICHE, P. (1950). 'A survey of weathering processes and products', *Univ. New Mexico, Publ. Geol.*, **3**,

RHODENBURG, H. (1969/70). 'Hangpedimentation und Klimawechsel als wichstigste Faktoren der Flächen und Stufenbildung in den wechseleuchten Tropen an Beispielen aus Westafrika, besonders aus dem Schichtstufenland Südost-Nigerias', *Göttinger Bodenkundliche Berichte*, **10**, 57–152. Also published in similar form under the same title in *Z. Geomorph.*, N.F., **14**, 58–78.

RICHARDS, P. W. (1957). 'Ecological notes on West African vegetation: I. The plant communities of the Idanre Hills, Nigeria', *J. Ecol.*, **45**, 563–577.

RODIER, J. (1961). *Régimes hydrologiques de l'Afrique Noire a l'ouest du Congo*, Paris.

ROE, F. W. (1951). *The geology and mineral resources of the Fraser's Hill area Selangor, Perak and Pehang, Federation of Malaya*, Geol. Surv. Federation Malaya, Mem. **5**.

ROUGERIE, G. (1955). 'Un mode de dégagement probable de certains dômes granitiques', *Compt. Rend. Acad. Sci.*, **246**, 327–329.

ROUGERIE, G. (1960). *Le façonnement actuel des modéles en Côte D'Ivoire forestière*, Mem. Inst. Francaise Afrique Noire, **58**, Dakar.

RUDDOCK, E. C. (1967). 'Residual soils of the Kumasi district in Ghana', *Geotechnique*, **17**, 359–377.

RUHE, R. V. (1956). *Landscape evolution in the High Ituri, Belgian Congo*, Publ. INEAC. Ser. Sci., **66**,

RUHE, R. V. (1960). 'Elements of the soil landscape', *7th Intern. Congr. Soil Sci.*, Madison, Trans. **4**, 165–170.

RUXTON, B. P. (1958). 'Weathering and sub-surface erosion in granite at the Piedmont Angle, Balos, Sudan', *Geol. Mag.* **95**, 353–377. ,

RUXTON, B. P. (1967). 'Slope wash under mature primary rainforest in Northern Papua', In: J. N. Jennings and J. A. Mabbutt (eds.), *Landform Studies from Australia and New Guinea*, University Press, Cambridge.

RUXTON, B. P. (1968a). 'Measures of the degree of chemical weathering of rocks', *J. Geol.*, **76**, 518–527.

RUXTON, B. P. (1968b). 'Rates of weathering of Quaternary volcanic ash in north-east Papua', *9th Intern. Congr. Soil Sci. Trans.*, **4**, 367–376.

RUXTON, B. P. (1968c). 'Order and disorder in landform', In: G. A. Stewart (ed.), *Land Evaluation*, Macmillan, Melbourne.

RUXTON, B. P. and BERRY, L. (1957). 'Weathering of granite and associated erosional features in Hong Kong', *Bull. geol. Soc. Am.*, **68**, 1263–1292.

RUXTON, B. P. and BERRY, L. (1959). 'The Basal Rock Surface on weathered granitic rocks', *Proc. Geologists Assoc.*, **70**, 285–290.

RUXTON, B. P. and BERRY, L. (1961a). 'Weathering profiles and geomorphic position on granite in two tropical regions', *Rev. Géomorph. Dyn.*, **12,** 16–31.

RUXTON, B. P. and BERRY, L. (1961b). 'Notes on facetted slopes, rock fans and domes on granite in east central Sudan', *Am. J. Sci.*, **259,** 194–206.

SANCHES FURTADO, A. F. A. (1968). 'Altération des granites dans les régions intertropicales sous différents climats', *9th Intern. Congr. Soil Sci.*, Adelaide, Trans. iv. 403–409.

SAPPER, K. (1935). 'Geomorphologie de feuchten Tropen', In: A. Hettner (ed.), *Geogr. Schriften*, **7,** 1–150.

SAVIGEAR, R. A. G. (1960). 'Slopes and hills in West Africa', *Z. Geomorph.*, Suppl. 1, Morphologie des versants, S156–171.

SAVIGEAR, R. A. G. (1965). 'A technique of morphological mapping', *Ass. Am. Geogr. Ann.*, **55,** 513–538.

SAVIGEAR, R. A. G. (1967). 'The analysis and classification of slope profile forms'. In: P. Macar (ed.), *L'Évolution des versants*, University of Liège, 271–290.

SCHALSCHA, E. B., APPELT, H. and SCHATZ, A. (1967). 'Chelation as a weathering mechanism 1: Effect of complexing agents on the solubilisation of iron from minerals and granodiorite', *Geochem. Cosmochim. Acta.*, **31,** (4), 587–596.

SCHATZ, A., SCHATZ, V. and MARTIN, J. J. (1957). 'Chelation as a biochemical weathering factor', *Bull. geol. Soc. Am.*, **68,** 1792–19.

SCHEIDEGGER, A. E. (1970). 'The large scale tectonic stress field of the earth', In: H. Johnson and B. L. Smith (eds.): *The Megatectonics of Continents and Oceans*, Rutgers University Press, New Brunswick, 223–240.

SCHUMM, S. A. (1956). 'Evolution of drainage systems and slope in badlands at Perth Amboy, New Jersey', *Bull. geol. Soc. Am.*, **67,** 597–646.

SCHUMM, S. A. (1963). *The disparity between present rates of denudation and orogeny*, U.S. geol. Surv., Professional Paper, 454-H.

SCHUMM, S. A. (1965). 'Quaternary paleohydrology', In: H. E. Wright and D. G. Frey (eds), *The Quaternary of the United States*, Princeton University.

SCHUMM, S. A. and LICHTY, R. W. (1965). 'Time, space and causality in geomorphology', *Am. J. Sci.*, **263,** 110–119.

SCHUYLENBORGH, J. van (1971). 'Weathering and soil forming processes in the tropics', In: *Soils and Tropical Weathering, Proc. Bandung Symp.*, UNESCO, 1969, 39–50.

SEGALEN, P. (1971). 'Metallic oxides and hydroxides in soils of the warm and humid areas of the world; formation, identification, evolution', In: *Soils and Tropical Weathering, Proc. Bandung Symp*, UNESCO, 1969, 25–38.

SELBY, M. (1967a). 'Morphometry of drainage basins in areas of pumice morphology', *Proc. Fifth New Zealand Geogr. Conf.*, 169–174.

SELBY, M. (1967b). 'Aspects of the geomorphology of the grewacke ranges bordering the lower and middle Waikato basins', *Earth Sci. J.*, **1**, (1), 1–22.

SHARPE, C. F. S. (1938). *Landslides and Related Phenomena*, Columbia University Press, New York.

SHERMAN, G. D. (1952). 'The genesis and morphology of the alumina-rich laterite clays, *Clay and Laterite Genesis, Am. Inst. Mining Metal*, 154–161.

SHORT, N. M. (1961). 'Geochemical variations in four residual soils', *J. Geol.*, **69**, 534–571.

SIMONETT, D. S. (1967). 'Landslide distribution and earthquakes in the Bewani and Torricelli Mountains, New Guinea', In: J. N. Jennings and J. A. Mabbutt (eds.), *Landform Studies from Australia and New Guinea*, University Press, Cambridge.

SINHA, N. K. P. (1966). *Geomorphic Evolution of Northern Rupununi, British Guiana* (unpublished Ph.D. thesis, McGill University, Toronto).

SIVAROJASINGHAM, S., ALEXANDER, L. T., CADY, J. G. and CLINE, M. G. (1962). 'Laterite', *Advan. Agron.*, **14**, 1–60.

SO, C. L. (1971). 'Mass movements associated with the rainstorm of June 1966 in Hong Kong', *Trans. Inst. Br. Geogr.*, **53**, 55–66.

SOMBRDEK, W. G. (1971). 'Ancient levels of plinthisation in N.W. Nigeria'. In: D. H. Yaalan (ed.), *Paleopedology*, Israel University Press, Jerusalem.

SPEIGHT, J. G. (1965). 'Flow and channel characteristics of the Angabunga River, Papua', *J. Hydrol.*, **3**, 16–36.

SPEIGHT, J. G. (1967). 'Geomorphology of Bougainville and Buka Islands', and Appendix 1: 'Explanation of the Land Systems descriptions', in C.S.I.R.O. Australia, *Lands of Bougainville and Buka Islands, Papua New Guinea*, Land. Res. Ser. **20.**

SPEIGHT, J. G. (1968). 'The parametric study of land form', in G. A. Stewart (ed.), *Land Evaluation*, Macmillan, Melbourne, 239–250.

SPEIGHT, J. G. (1971). 'Long-normality of slope distributions'. *Z. Geomorph.*, N.F., **14**, (3), 290–311.

STARKEL, L. (1966). 'Post-glacial climate and the moulding of European relief', *Royal Met. Soc. Symp. on World Climate 8000–0 B.C.*, 15–33.

STEPHENS, C. G. (1946). *Pedogenesis following the dissection of laterite regions in southern Australia*, Council Sci. Ind. Res. Australia, Bull., **206.**

STEPHENS, C. G. (1967). 'Soil stratigraphy and its applications to correlation of Quaternary deposits and landforms and to soil science – A review of Australian experience'. In: R. B. Morrison and H. E. Wright (eds.), *Quaternary Soils*, Proc. Vol. 9, CII Congress INQUA.

STEPHENS, C. G. (1971). 'Laterite and silcrete in Australia etc.', *Geoderma*, **5**, 5–52.

STEPHENS, R. E. and CARRON, M. K. (1948). 'Simple field test for distinguishing minerals by abrasion pH', *Am. Mineralogist*, **33**, 31–49.

STODDART, D. R. (1969). 'Climatic geomorphology: review and reassessment', In: C. Board *et al* (eds.), *Progress in Geography*, **1**, Arnold, London.

STRAHLER, A. N. (1952). 'Hypsometric (area–altitude) analysis of erosional topography', *Bull. geol. Soc. Am.*, **63**, 1117–1142.

STRAHLER, A. N. (1952a). 'Dynamic basis of geomorphology', *Bull. geol. Soc. Am.* **63**, 923–38.

STRAKHOV, N. M. (1967). *Principles of Lithogenesis*, Vol. **1**, Oliver and Boyd, Edinburgh.

SWAN, S. B. St. C. (1967). 'Maps of two indices of terrain, Johor, Malaya', *J. Trop. Geog.*, **25**, 48–57.

SWAN, S. B. St. C. (1970a). 'Piedmont slope studies in a humid tropical region, Johor, Southern Malaya', *Z. Geomorph.*, Suppl., **10**, 30–39.

SWAN, S. B. St. C. (1970b), 'Analysis of residual terrain, Johor, Southern, Malaya', *Ass. Am. Geogr. Ann.*, **60**, 124–133.

SWAN, S. B. St. C. (1970c). 'Land surface mapping, Johor, West Malaysia', *J. Trop. Geog.* **31**, 91–103.

SWAN, S. B. St. C. (1972). Land surface evolution and related problems with reference to a humid tropical region: Johor, West Malaysia, *Z. Geomorph. N.F.*, **16**, 160–181.

SWEETING, M. M. (1958). 'The karstlands of Jamaica', *Geograph. J.*, **124**, 184–199.

SWEETING, M. M. (1968). 'Karst', In: R. W. Fairbridge (ed.), *Encyclopaedia of Geomorphology*, Reinhold, New YoFk, 582–587.

TATOR, B. A. (1952). 'The climatic factor and pediplanation', *Congr. Geol. Intern., Compt. Rend. 19e*, Algiers, 1952, **7**, 121–130.

TATOR, B. A. (1952–3). 'Pediment characteristics and terminology', *Assoc. Am. Geogr. Ann.*, **42**, 295–317; **43**, 47–53.

TE PUNGA, M. T. (1964), 'Relict red-weathered regolith at Wellington', *New Zealand J. geol. Geophys.*, **7**, (2), 314–339.

TERZAGHI, K. (1950). 'The mechanism of landslides', *Bull. geol. Soc. Am.*, **83**, 123.

TERZAGHI, K. (1962). Dam foundations on sheeted granite. *Geotechnique*, **12**, 199–208.

THOMAS, M. F. (1965a). 'Some aspects of the geomorphology of domes and tors in Nigeria', *Z. Geomorphol.*, N.F., **9**, 63–81.

THOMAS, M. F. (1965b). 'An approach to some problems of landform analysis in tropical environments', In: P. D. Wood and J. B. Whittow (eds.), *Essays in Geography for Austin Miller*, University of Reading.

THOMAS, M. F. (1966a). 'Some geomorphological implications of deep weathering patterns in crystalline rocks in Nigeria', *Trans. Inst. Br. Geogr.*, **40**, 173–193.

THOMAS, M. F. (1966b). 'The origin of bornhardts', *Z. Geomorph.*, N.F., **10**, 478–480.

THOMAS, M. F. (1967). 'A bondhardt dome in the plains near Oyo, western Nigeria', *Z. Geomorph., N.F.*, **11**, 239–261.

THOMAS, M. F. (1968). 'Some outstanding problems in the interpretation of the geomorphology of tropical shields', *Br. Geomorph. Res. Gp. Publ.*, **5**, 41–49.

THOMAS, M. F. (1969). 'Geomorphology and land classification in tropical Africa', In: M. F. Thomas and G. Whittington (eds.), *Environment and Land Use in Africa*, Methuen, London.

THOMAS, M. F. (1974). Granite landforms, a review of some recurrent problems of interpretation.' In: E. H. Brown and R. S. Waters (eds.), *Instr. Br. Geogr. Spec. Publ.*, 7.

THORBECKE, F. (1951). 'Im Hochland von Mittel-Kamerun – Physische Geographie des Ost-Mbamlands', *Universität Hamburg Abhandlungen aus dem Gebiet der Auslandskunde*, **4.**

THORNBURY, W. D. (1954). *Principles of Geomorphology*, Wiley, New YoFk.

THORP, M. B. (1967a). 'Closed basins in Younger Granite Massifs, northern Nigeria', *Z. Geomorph.*, **11,** N.F., 459–480.

THORP, M. B. (1967b), 'Jointing patterns and the evolution of landforms in the Jarawa granite massif, northern Nigeria', In: R. Lawton and R. W. Steel (eds.), *Essays in Geography*, Longmans, London.

THORP, M. D. (1970). 'Landforms', In: M. J. Mortimore, *Zaria and its Region*, Ahmadu Bello University, Zaria, Dept. Geogr. Occ. Paper, **4,** 13–32.

TRENDALL, A. F. (1962). 'The formation of "Apparent Peneplains" by a process of combined lateritisation and surface wash', *Z. Geomorph.*, N.F., **6,** 183–197.

TRICART, J. (1972). *The Landforms of the Humid Tropics, Forests and Savannas*, Transl. from the French by C. J. K. de Jonge, Longmans, London.

TUAN, Yi-Fu. (1959). *Pediments in southeastern Arizona,* Univ. California Publ. Geograph., **13, .**

TWIDALE, C. R. (1962). 'Steepened margins of inselbergs from northwestern Eyre Peninsula, South Australia', *Z. Geomorph.*, N.F., **6,** 51–69.

TWIDALE, C. R. (1964). 'A contribution to the general theory of domed inselbergs', *Trans. Inst. Br. Geogr.*, **34,** 91–113.

TWIDALE, C. R. (1967). 'Hillslopes and pediments in the Flinders Ranges, South Australia', In: J. N. Jennings and J. A. Mabbutt (eds.), *Landform studies from Australia and New Guinea*, University Press, Cambridge.

TWIDALE, C. R. (1968). 'Pediments', In: R. W. Fairbridge (ed.), *The Encyclopedia of Geomorphology*, Reinhold, New YoFk, 817–818.

TWIDALE, C. R. and CORBIN, M. (1963). 'Gnammas', *Rev. Geomorph. Dyn.*, **14,** 1–20.

TWIDALE, C. R. (1971). *Structural Landforms*, M.I.T. Press, Cambridge, Mass.

USSELMAN, P. (1968). 'Discussion of M. R. M. de Meis and J. X. da Silva: Movement de masse récentes a Rio de Janeiro—Une étude de géomorphologie dynamique', *Rev. Géomorph. Dyn.*, **18,** (4), 1–15.

VAN DIJK, D. C., RIDDLER, A. M. H. and ROWE, R. K. (1968). 'Criteria and problems in ground surface correlations with reference to a regional correlation in south eastern Australia', *9th Intern. Congr. Soil Sci.*, Adelaide, Trans. **4**, 131–137.

VANN, J. H. (1963). 'Developmental processes in laterite terrains in Amampa', *Geogr. Rev.*, **53**, 406–417.

VAN ZINDEREN BAKKER, E. M. (1967). 'Upper Pleistocene and Holocene stratigraphy and ecology on the basis of vegetation changes in sub-Saharan Africa', In: W. W. Bishop and J. D. Clark (eds.), *Background to Evolution in Africa*, University of Chicago Press, Chicago.

VOGT, J. (1966). 'Le complex de la "Stone-line"—Mise en point', *Bull. Bur. Récherche Géol. Min.*, **3**, 49.

WAHRHAFTIG, C. (1965). 'Stepped topography of the southern Sierra Nevada, California', *Bull. geol. Soc. Am.*, **76**, 1165–1190.

WALKER, D. (1970). 'The changing vegetation of the montane tropics', *Search*, **1**, (5), 217–221.

WAMBEKE, A. R. Van (1962). 'Criteria for classifying tropical soils by age', *J. Soil Sci.*, **13**, (1), 124–132.

WARD, W. T. (1951). 'The tors of central Otago', *New Zealand J. Sci. Tech.*, B, **54**, 191–200.

WATERS, R. S. (1952). 'Pseudo-bedding in the Dartmoor granite', *Roy. Soc. Cornwall. Trans.*, **18**, 456–462.

WATERS, R. S. (1957). 'Differential weathering and erosion on oldlands', *Geogr. J.*, **123**, 510–509.

WATSON, J. P. (1964). 'A soil catena on granite in Southern Rhodesia', *J. Soil Sci.*, **15**, 238–257.

WAYLAND, E. J. (1934). 'Peneplains and some other erosional platforms', *Bull. Geol. Surv. Uganda, Annual Rept.*, Notes 1, **74**, 366.

WAYLAND, E. J. (1947). 'The study of past climates in tropical Africa', *Proc. Pan African Congr. Prehistory*, 59–66.

WEBSTER, R. (1965). 'A catena of soils on the Northern Rhodesia plateau', *J. Soil Sci.*, **16**, (1), 31–43.

WELLMAN, H. W. and WILSON, A. T. (1965). 'Salt weathering, a neglected erosive agent in coastal and arid environments', *Nature*, London, **205**, 1097–1098.

WENTWORTH, C. K. (1943). 'Soil avalanches on Oahu, Hawaii', *Bull. geol. Soc. Am.*, **54**, 53–64.

WEST, G. and DUMBLETON, J. J. (1970). 'The mineralogy of tropical weathering illustrated by some West Malaysian soils', *Quat. J. Eng. Geol.*, **3**, (1), 25–40.

WHITE, W. A. (1944). 'Geomorphic effects of indurated veneers on granite in the south eastern States', *J. Geol.*, **52**, 333–341.

WHITE, W. A. (1945). 'The origin of granite domes in the south east piedmont', *J. Geol.*, **53**, 276–282.

WHITE, W. A. (1972). 'Deep erosion by continental ice-sheets', *Bull. geol. Soc. Am.*, **83**, 1037–1056.

WIGWE, G. A. (1966). *Drainage composition and valley forms in parts of Northern and Western Nigeria* (unpublished Ph.D. thesis, University of Ibadan).

WILHELMY, H. (1958). *Klimamorphologie der. Massengesteine,* Westerman, Brunswick.

WILLIAMS, M. A. J. (1968). 'Termites and soil development near Brocks Creek, Northern Territory', *Aust. J. Sci.,* **31,** 153–154.

WILLIAMS, M. A. J. (1969). 'Prediction of rainsplash erosion in the seasonally wet tropics', *Nature,* **222,** (No. 5195), 763–765.

WILLIAMS, P. W. (1971). 'Illustrating morphometric analysis of Karst with examples from New Guinea', *Z. Geomorph.*, N.F., **15,** (1), 40–61.

WILLIS, B. (1936). *Studies in Comparative Seismology: East African Plateaux and Rift Valleys,* Carnegie Inst., Washington.

WILSON, N. W. and MARMO, V. (1958). 'Geology, Geomorphology and Mineral Resources of the Sula Mountains', *Bull Geol. Surv. Sierra Leone,* **1.**

WOLMAN, M. G. and MILLER, J. P. (1960). 'Magnitude and frequency of forces in geomorphic processes', *J. Geol.,* **68,** 54–74.

WOOLNOUGH, W. G. (1918). 'Physiographic significance of laterite in Western Australia', *Geol. Mag. N.S.,* **6,** (5), 385–393.

WOOLNOUGH, W. G. (1927). 'The duricrust of Australia', *J. Proc. Roy. Soc. New South Wales,* **61,** 25–53.

WOOLNOUGH, W. G. (1930). 'Influence of climate and topography in the products of weathering', *Geol. Mag.,* **67,** 123–132.

WRIGHT, R. L. (1963). 'Deep weathering and erosion surfaces in the Daly River Basin, Northern Territory', *J. geol. Soc. Aust.,* **10,** 151–163 .

YOUNG, A. (1963). 'Soil movement on slopes', *Nature*, 200, (No. 4902), 129–130.

YOUNG, A. (1969). 'Natural resource survey in Malawi: some considerations of the regional method in environmental description', In: M. F. Thomas and G. Whittington (eds), *Environment and Land Use in Africa*, Methuen, London.

YOUNG, A. (1970). 'Slope forms in the Xarantina-Cachimbo area', In: *Geographical Research on the Royal Society/Royal Geographical Society's Expedition to north-eastern Mato Grosso, Brazil: A Symposium, Geogr. J.,* **136,** 383–392.

YOUNG, A. (1972). '*Slopes*', Oliver and Boyd, Edinburgh.

YOUNG, A. (1973). Personal Communication.

Subject Index

Page numbers printed in bold refer to illustrations

Author Index

Page numbers in bold refer to illustrations

Geographical Index

Page numbers in bold refer to illustrations